Projects for the Car and Garage

Macmillan Electronic Projects Series

Audio Circuits and Projects (revised edition) Graham Bishop
Program and Electronic Projects for the BBC, Electron and Spectrum Computers Graham Bishop
Projects for the Car and Garage (revised edition) Graham Bishop
Cost-effective Electronic Construction John Watson

Projects for the Car and Garage

Graham Bishop

Revised Edition

© Graham Bishop 1980, 1985

All rights reserved. No reproduction, copy or transmission of this publication may be made without written permission.

No paragraph of this publication may be reproduced, copied or transmitted save with written permission or in accordance with the provisions of the Copyright Act 1956 (as amended), or under the terms of any licence permitting limited copying issued by the Copyright Licensing Agency, 33–4 Alfred Place, London WC1E 7DP.

Any person who does any unauthorised act in relation to this publication may be liable to criminal prosecution and civil claims for damages.

First edition (under the title
Electronic Projects 2 — Projects for
the Car and Garage), 1980
Revised edition 1985
Reprinted 1988, 1989

Published by
MACMILLAN EDUCATION LTD
Houndmills, Basingstoke, Hampshire RG21 2XS
and London
Companies and representatives
throughout the world

Printed in Hong Kong

British Library Cataloguing in Publication Data

Bishop, G. D.
    Projects for the car and garage. —Rev. ed.
    —(Electronic projects)
    1. Automobiles—Electronic equipment
    I. Title   II. Series
    629.2'7   TL272.5

ISBN 0–333–37220–4

Contents

Preface		viii
Preface to the Revised Edition		ix
General Instructions		x

1 Ignition Circuits   1

     1.1   Conventional Ignition   2
     1.2   Ignition Problems   4
     1.3   dc/dc Converter Circuit   5
     1.4   Electronic Ignition   8
     1.5   Electronic Ignition Circuit   11
            1.5.1   Construction Sequence   13
     1.6   CD Ignition Installation   15
     1.7   Ignition Timing Light   16
     1.8   A Dwell Meter   17
     1.9   Ignition Suppression   19
     1.10   Ignition Booster Circuit   23

2 Car Theft Circuits   25

     2.1   Immobilising Circuits   25
     2.2   Alarm Circuits   26

3 Lighting Circuits — 29

- 3.1 Interior Light Extender — 30
- 3.2 Nightlight — 32
- 3.3 'Lights-on' Reminder — 32
- 3.4 Alarm Circuit — 34
- 3.5 'Ignition Key In' Reminder — 36
- 3.6 Light Monitors (Fibre-optic) — 36
- 3.7 Electrical Car Light Failure Monitor — 39
- 3.8 Audible Turning Indicator — 40
- 3.9 Emergency Flasher — 41
- 3.10 Trailer Flashers — 42

4 Accessory Circuits — 44

- 4.1 Car Radio Audio Booster — 44
  - 4.1.1 Stereo Systems — 49
  - 4.1.2 Pseudo-stereo Systems — 51
- 4.2 Car Aerial Amplifier — 55
- 4.3 Rev Counter — 55
- 4.4 Car Clock — 58
- 4.5 Wiper Delay Circuit — 60
- 4.6 Emergency Beacon — 63
- 4.7 Car Appliance Supplies — 63
- 4.8 Ammeter and Voltmeter Connections — 68
- 4.9 Battery Condition Tester — 69
- 4.10 Alternator Regulator Circuit — 72

5 Garage and Test Gear — 75

- 5.1 Battery Charger — 75
  - 5.1.1 Safety Precautions — 78
  - 5.1.2 Layout — 79
  - 5.1.3 Testing — 79
- 5.2 Ice Warning Device — 81
- 5.3 Car Ice Alarm — 83
- 5.4 Gauge Alarm Circuits — 83
- 5.5 Parking Meter Reminder — 83
- 5.6 Accelerometer — 86
  - 5.6.1 Setting Up — 88

Appendix I Components Required for Projects — 90
Appendix II Specifications of Semiconductor Devices — 103
Appendix III Resistor/Capacitor Colour Coding — 106

Index — 113

Preface

This book is one of a series of electronics hobbies books written for the relative newcomer to electronics construction, and also for the experienced constructor. This book includes about 50 project circuits and ideas which are all **fully tested** and which use readily obtainable components. The projects use, where necessary, printed circuit board from Vero Electronics Ltd which needs **no track cutting**, uses 0.1 in. spacing throughout, and will accept most components without the need for extra hole drilling. The copper-covered board is soft enough to be cut by the bluntest saw, although the strength is perfectly adequate for the applications given. Connection from one track to its neighbour is shown in each circuit by

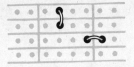

which is carried out by holding the soldering iron between the two tracks until the solder runs across the tracks. The tracks can be linked right across the board using this method, each joint taking less than 1 second!

The accent in this book is on **safety**, since a car travelling at 70 mph on the motorway must be 100 per cent safe and reliable despite, or perhaps because of, the addition of electronic circuits. Constant references are made to the dangers of the car battery and to fusing and protection of circuits, for good reason.

Several of these circuits can be constructed on a single Veroboard; they can be mounted for convenience all on one panel near the dashboard of the car with the necessary meter(s) and lights/alarms all neatly fixed in place. This entire board can then be connected to the 12 V supply via one fuse from the ignition switch.

These circuits gave me an immense amount of enjoyment to design and make, I hope that you enjoy making them and using them.

GRAHAM BISHOP

Preface to the Revised Edition

Surprisingly, there have been few developments in car electronics since the publication of the first edition. The car computer and microprocessor can now be found in a few select cars such as the BL Maestro, together with 'talking chips'. These are extremely complex and unfortunately cannot be adapted for hobbyist use; devices such as these can only be built into a new car, not added on to an existing model.

1984   GRAHAM BISHOP

General Instructions

Important: Read these notes before installing any circuit into your car.

A car electrical system consists of a 12 V lead–acid battery and the following.

(1) The **ignition circuit** including distributor, ignition coil, spark-plugs.
(2) The **lighting circuits** including flasher unit.
(3) **Battery-charging circuit** including dynamo/alternator and regulator.
(4) **Instrument circuits** incorporating gauges, warning lights and displays.
(5) **Auxiliary circuits** including car radio, cassette player, and so on.

The circuits described in this book are grouped into all of these areas of the car electrical system but before proceeding with installation you are advised to obtain a copy of the complete wiring diagram of the car involved, which will tell you

(1) The polarity of the earth. All circuits in this book assume a **negative earth** since this is the most common; similarly all circuits

assume a 12 V system although some 6 V cars do exist. Conversion of the circuits in this book to a car with a positive earth presents no problem, although the switching and insulation from the chassis might require some thought; in many circuits *npn* and *pnp* transistors can be interchanged accordingly. 6 V operation does present a problem, and many of the given circuits are not suitable.

(2) The method of switching the accessories, whether the switch is in the positive or the negative lead, so that additional circuits can be connected in parallel.

(3) The type of ignition system used and whether dynamo or alternator driven. This information is needed for the installation of electronic ignition.

(4) The positions of the fuses and a convenient source of +12 V.

(5) The rating of all fuses and lights.

Overloading of the car circuits can cause fuses to blow at inconvenient times, both when the added electronic circuit is working normally and when it is faulty. It is best to fuse any added circuits separately or install an electro-mechanical cut-out in series with the +12 V supply.

One or two additional points to remember with car electronics are as follows.

(1) Always remove the battery 'earth' lead from the battery while installing a circuit, in particular circuits that tap the 12 V supply direct from the battery. If the positive lead accidentally shorts to chassis up to 100 A can flow causing a fire or severe burns!

(2) Many low voltage car appliances such as the motors and lights carry several amps along their supply leads and so any additional circuits and their connecting leads should also be capable of carrying these currents without overheating.

(3) Many spurious voltage pulses travel around a car electrical system and so good decoupling of the power input leads to any electronic circuit is essential using a 100 μF capacitor or thereabouts.

(4) Avoid having circuits running while you are away from the car except anti-theft alarms. Any fault that develops can cause trouble or drain the battery.

(5) If you alter the car wiring at all by adding electronic circuits do **not** go to a garage complaining of electrical troubles with your car. The mechanic will quickly cut out all your good work and restore your car to its normal 'standard' condition before finding out what is wrong.

(6) Make your circuit layouts as neat as possible. This simplifies

servicing and increases reliability. It is good practice always to adopt a consistent colour scheme for supply leads (red, black, for example) and inputs/outputs, and to label all controls and wires for future reference. Label the sockets too.

A full list of components is included at the end of the book. It will be noticed that very few different types of transistor and diode are used: the BC109 and BC177 are used for general amplification with the BD131/711 and 2N3055 for power output. A list of near equivalents is also included. All resistors are $\frac{1}{8}$ W and all capacitors are low voltage unless otherwise stated. When integrated circuits are used it is advisable to use an IC socket rather than solder the IC directly into the circuit and risk damage by overheating. The IC can then be removed, if necessary, in one piece. Watch the pin connections of some devices such as ICs and thyristors, for near equivalents often use different pin connections.

The accent has been on simplicity, low cost and reliability in these circuits. Not one should be beyond the capabilities of anyone who has built one or two electronics projects before.

Publisher's Note

While every effort has been made to ensure the accuracy of the projects and circuits in this book, neither the Publisher nor the author accepts liability for any injury or loss resulting from the construction of any of the designs published herein.

The Publisher will, however, be pleased to hear from readers who have corrections to the text or genuine queries, and will refer any such queries to the author.

1
Ignition Circuits

No book on electronic circuits for cars would be complete without an electronic ignition circuit to replace the conventional inductance-discharge system with a **capacitance-discharge** circuit that will prove to be much more reliable and efficient on cold wintry mornings. The fuel consumption will also be improved because the timing of the

Some of the projects, photographed against a car bonnet to show the scale.

sparks to the spark-plugs is more exact (provided the timing is adjusted correctly, as well as the spark-gaps of the contact breakers and spark-plugs). This chapter first briefly describes the conventional system used in most modern cars which do not have electronic ignition already! A trustworthy dc/dc converter circuit, which has uses other than for ignition, is then described and the chapter also includes notes on ignition radio interference suppression and a circuit of a dwell meter used to adjust the contact breaker gap.

I must start by pointing out that every make of car has its own systems and characteristics and the circuit given must be adapted accordingly. I have tried out these circuits on a Mini and a large Ford car with no problems — but the car wiring diagram must be consulted before tackling these ignition projects, and the entire job planned out carefully.

1.1 Conventional Ignition

The system is outlined in figure 1.1. The ignition circuit is essential to the working of the car engine. The engine is a highly tuned machine which depends on ignition of the petrol/air mixture at the point when the piston is almost at the top of its stroke, thereby igniting the fuel and explosively forcing the piston to turn the engine. Four or six cylinders require four or six ignition circuits, all precisely timed.

The ignition is created by a high voltage spark which is struck across the spark-plug gap of about 25 thousandths of an inch (25 thou) and is generated in an ignition coil with a primary and a secondary winding. The turns ratio within this coil is of the order of 10 000 to 1 and so a 12 V pulse applied to the primary winding is transformed up to 120 000 V in the secondary. If this high voltage is applied to the spark-plug, a spark will ignite the mixture and produce power in the cylinder.

In order to fire each cylinder in sequence, a distributor switches the high voltage from cylinder to cylinder and, at the same time, operates a small switch to generate the 12 V pulses in the primary winding. This switch is the contact breaker whose contacts are opened and closed by a cam connected to the shaft of the distributor. The contact breaker is adjusted very carefully so that the gap is correct (25 thou) and the timing of the opening of the contacts is also correct; this is called engine tuning. The contact breaker points switch 10 million or so times every 1000 miles and so tend to require regular maintenance. It is for this reason that electronic ignition or 'contactless' contact breakers are preferred.

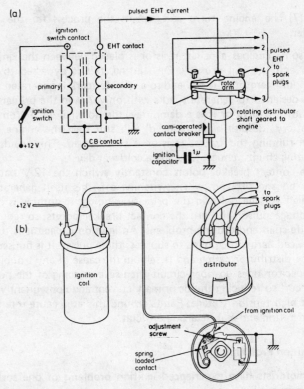

Figure 1.1 Conventional ignition. (a) Conventional circuit. (b) Hardware: the contacts open four times per revolution to activate EHT spark in corresponding spark-plug, the spark being routed to appropriate plug by conducting brass rotor arm at distributor housing.

To summarise:

(1) With the ignition switch on, the contact breaker points are closed.

(2) Current from the car battery passes through the ignition coil primary.

(3) A magnetic field is produced in the primary.

(4) As the starter turns the engine, one piston rises and the petrol/air mixture is compressed. At the same time the distributor shaft turns and opens the contact breaker points, at top dead centre.

(5) The magnetic field in the coil collapses and a high voltage spark is generated in the secondary, and thence to the spark-plug at the top of the cylinder.

(6) The mixture ignites and power is produced.

(7) The engine turns and repeats the process for the next cylinder.

In some engines a ballast resistor is placed between the ignition switch and the coil primary. On starting, this is bypassed to give maximum energy to the coil and so a reliable spark is generated. The coil is designed to generate sparks with only 7 V in the primary. If the leads are dirty or it is a damp day, the nominal 12 V from the car battery (despite being reduced) will still start the engine. For normal running this ballast resistor is switched in. This avoids the frustrating chug...chug...chug on a cold wet day.

The contact breaker points constantly switch the 12 V battery supply on and off and, when switching off, this itself generates a very high voltage pulse in the primary winding. If nothing is done this voltage pulse will burn the contact breaker points, so reducing their life span and causing problems. An ignition capacitor is therefore placed across the points to suppress these pulses; it is housed inside the distributor casing and is seldom the cause of any trouble. If the capacitor does go open-circuit then severe arcing of the points will occur, so reducing the dc primary current and consequently the output high tension voltage. Faults around this area cause intermittent operation and make starting difficult.

1.2 Ignition Problems

Most motorists have experienced ignition problems of one sort or another caused by

(1) A poorly kept battery which struggles to turn the engine thereby using all battery power for the starter. This leaves insufficient reserves for the ignition circuit.

(2) A poorly set contact breaker gap and plug gaps.

(3) Damaged plugs with possible internal short-circuit or low resistance (oily dirt causes this).

(4) Deposits on the contact breaker points and/or the plugs and/or the connectors in the distributor top. (Careful cleaning or renewal is the normal cure.)

(5) Condensation on the entire system; this partly short-circuits the EHT passing along the high tension leads to chassis. A spectacular experiment may be carried out on a dark morning in your garage after a heavy dew has fallen on the car and engine. The entire ignition system lights up blue owing to the many EHT paths both inside the EHT leads and outside — provided you can get the engine to start!

These ignition problems may be cured by

(1) Buying a new battery or improving its condition.

(2) Buying a new set of plugs and points and setting their gaps correctly.

(3) Cleaning all connectors and leads, particularly in the EHT circuit.

(4) Spraying a damp engine with WD40 to repel the moisture (this never seems to work with me!).

(5) Boosting the EHT to a much higher level to increase the energy of the spark at the plugs and cause the engine to start.

If the costs of these remedies are analysed, (1) and (2) are relatively low cost when you consider the likely time involved in tracing ignition faults; the purchase of a battery, plugs and points is thoroughly recommended as a first step. The cleaning of leads and connectors is quick and greatly improves performance at no cost; (4) is hardly worth it, whereas (5) involves the installation of electronic ignition, most commercial systems costing about the same as a new battery.

1.3 dc/dc Converter Circuit

The basis of the capacitor-ignition circuit described in this chapter lies in the dc to dc converter circuit of figure 1.2, which has many other uses in the car and garage. This circuit converts the 12 V car battery supply into a supply of several hundred volts for uses to be described later. Two outputs are provided, one is dc (for use in the ignition circuit) and the other is ac (for use as a substitute for mains

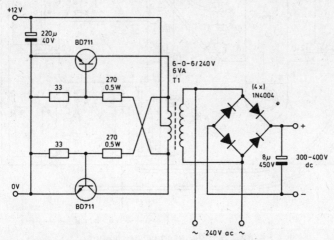

Figure 1.2   dc/dc converter. Basic converter circuit.

6 / Projects for the Car and Garage

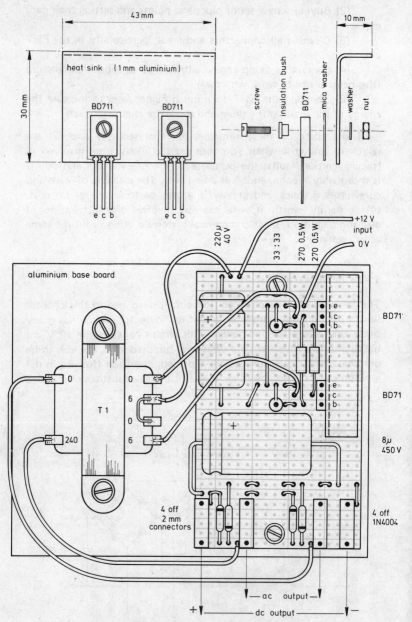

Figure 1.3   dc/dc converter layout.

voltage for small appliances such as shavers, camping lamps, televisions, and so on). The converter circuit is completely self-contained and built on to a baseboard of only 100 mm x 100 mm.

Figure 1.2 shows this simple **converter** circuit whose layout is seen in figure 1.3 on a baseboard only 100 mm x 100 mm. If housed in a metal box with the ignition circuit, the transformer and circuit board can be screwed directly to the box. The converter principle involves the generation of a squarewave from the low voltage 12 V supply using a multivibrator, and the stepping up of this ac output to 240 V using a 6–0–6/240 V transformer. The 300 to 400 V is obtained by rectifying the ac waveform with a bridge rectifier system. There is a smoothing capacitor, of 8 μF 450 V. It is difficult to state the exact ac and dc output levels of this circuit since this depends on the components used, the rating of the transformer and the load that is connected; the off-load dc output is about 400 V.

The output voltage will fall considerably if a current in excess of about 0.5 A is drawn from the converter; this is due to the relatively poor **regulation** properties of the circuit. This is unimportant when used in the ignition circuit since the output currents are very small. When used to run a 'mains' ac appliance, however, the current rating of the appliance will determine the resultant output voltage. For example a 240 V 100 W lamp bulb will draw about 400 mA and the converter output voltage will probably fall to around 220 V provided the converter components are rated at the higher current levels (2N3055s, 10 A transformer, and so on). Remember that a higher output voltage and current from the secondary of T_1 corresponds to a far greater current into the primary of T_1. The 100 W output referred to above necessitates about 125 W of power into the primary winding, assuming 25 W are wasted in the transformer. Since this is a 6–0–6 transformer then **10 A** will flow into T_1 from the 2N3055s. If you want to use the circuit for lighting, a **fluorescent** lamp with equivalent light output to a 100 W lamp will consume far less current and so is preferred.

If you intend to run your mains television receiver from this circuit, a **very large** transformer will be necessary and it would be cheaper for you to invest in a mains/battery television receiver which is designed to be run directly from a 12 V car battery! If such an appliance is run from this converter circuit, beware of initial overloading of the appliance because the off-load voltage of 300 to 400 V falls to 240 V or thereabouts on load. This initial surge can cause problems and so you are advised to connect the appliance to the converter with both switched off. **Then** switch on the appliance and **finally** switch on the converter whose output voltage will rise from zero to 240 V, instead of falling from 400 to 240 V.

The output frequency of the converter depends on the transistors used, the BD711 transistors give a frequency of around 100 Hz, 2N3055s produce about 300 Hz. This is due to the different base voltages produced as a result of the different base currents, which alter the trigger timing of the multivibrator and so the frequency For most applications (lights, ignition, ignition timing and so on) this is unimportant but for some appliances the high frequency may upset the performance. Any device which is dependent on 'mains' frequency will be driven silly by the increase to 300 Hz or 1 kHz and damage can result. Devices with an input 'mains' transformer and power supply smoothing circuit — such as audio amplifiers and tape recorders — will not object to the high frequency although there will be a change to the output characteristics of the power supply. A record player turntable or any device using a synchronous motor will follow the frequency of the input 'mains' and will run much too fast, although there will be no damage. You are advised to consult the manufacturer's literature to see if input frequency is important. However, this is a handy circuit to have around when winter power cuts occur.

1.4 Electronic Ignition

The system used is termed **capacitor-discharge ignition**, outlined in figure 1.4. The first point to notice is that the **existing** ignition coil and contact breakers are used. All that is done is to replace the original 12 V switching circuit in the primary winding by a 300 or 400 V circuit. This converts a current circuit which is upset by lead and stray resistance into a voltage circuit, which has much greater efficiency.

The **dc/dc converter** generates the high voltage dc which is applied to the ignition coil primary via a 0.47 μF capacitor. This voltage must be pulsed into the ignition circuit and so a thyristor (SCR) is placed across the converter output, to ground the output except for small intervals of time when the contact breaker points open. The contact breakers, unfortunately, do not open cleanly and a **damped oscillation** (illustrated in figure 1.8 where this is used to 'drive' the dwell meter) is normally obtained. A **bounce suppression** circuit is added therefore to clean up the contact breaker pulse for firing the thyristor just **once** when required. The thyristor trigger is obtained from a positive pulse created as the CB contacts open. When closed the gate is at zero potential and the thyristor is off. Meanwhile the 0.47 μF capacitor is fully charged up to 300 to 400 V.

When the CB contacts open, the gate receives a 12 V pulse to turn it on, causing the thyristor to conduct and ground the left plate of the capacitor. The capacitor charge is now released into the ignition

Ignition Circuits / 9

coil primary to be amplified by the coil and create 40 kV or so EHT. The contact breakers are now isolated from the EHT circuit and the contacts keep very clean, a 500 mA current being adequate for this purpose. The thyristor will remain switched on until the anode voltage falls below that of the cathode. This is achieved by causing the 0.47 µF capacitor to resonate with the coil at a frequency whose period is short. The thyristor anode voltage falls below that of the cathode after the first half-cycle of oscillation and — in this

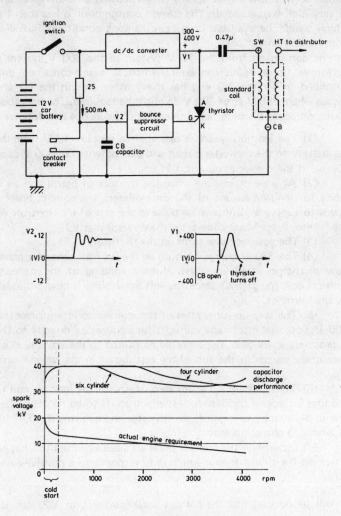

Figure 1.4   Capacitor-discharge ignition.

circuit — switches off after about 150 µs from switching on, to enable the capacitor again to charge up, even though the contact breaker points may still be open. The waveform is illustrated in figure 1.4, together with the EHT variation with engine speed.

This circuit **eliminates arcing** at the points and therefore any build-up of deposit; **plug life is increased** owing to the faster generation of EHT; **contact breaker bounce is corrected** and the **EHT is maintained** at very **low** engine speeds (when starting a cold engine) and at very **high** engine speeds. The current consumption is less than 1 A which would be available under **any** running conditions including poor battery conditions.

The conversion from the 12 V system to the 300 V system is performed in a **dc/dc converter**, which is a self-contained unit described in section 1.3 and has many other uses in the car and garage where 250 V ac or 300 V dc might be required. The operation of the capacitor-discharge ignition circuit of figure 1.4 is as follows.

(1) The ignition switch is operated to supply +12 V from the car battery to the converter circuit and a dc potential of 300 V or so appears at the converter output, V1.

(2) At a point in time when the mixture of petrol and air is ready for ignition in one of the car cylinders, the contact breaker **opens** to supply a 12 V positive pulse to the gate of the thyristor via the bounce suppression circuit, see the waveform at V2.

(3) The positive pulse turns **on** the thyristor.

(4) The 300 V charge built up on the 0.47 µF capacitor plates now discharges itself into the primary winding of the standard ignition coil (point SW) since the left-hand plate is now grounded by the thyristor.

(5) The step-up turns ratio of the ignition coil transforms this 300 V potential into many kilovolts for application directly to the appropriate spark-plug, the spark being routed to the spark-plug by the rotary switch in the top of the distributor in the conventional way.

(6) The 0.47 µF capacitor resonates with the ignition coil to produce a damped oscillation, V1, which starts to go negative after one cycle thereby turning off the thyristor and allowing the 0.47 µF capacitor to charge up again.

(7) Meanwhile the contact breakers close again but this has no effect on the thyristor gate which only responds to a positive-going pulse.

It will be noticed that the contact breaker waveform, V2, also has damped oscillation superimposed on it, and it is the function of the bounce suppression circuit to minimise this and ensure that the thyristor gate receives only one initial pulse and not a series of smaller pulses. The existing contact breaker timing capacitor is left in position since this ensures perfect synchronisation between the opening of the points, the firing of the thyristor and the timing of the spark, as was the case with the conventional ignition system. It is unfortunate that the presence of this timing capacitor causes oscillation, but the bounce suppression circuit of just two components minimises this very efficiently without changing the delay time of 125 μs or so necessary for accurate timing. It will also be evident that the contact breaker points no longer pass the high voltages of the original system, they merely switch on and off a 500 mA, 12 V signal to switch the thyristor gate. The points do not arc or corrode and their life is greatly increased. The big advantage of this circuit lies in the fact that the original components are used and very little modification to the car electrics is necessary. The **black box** that is added in the next section can be plugged or switched over from conventional to CD in a few seconds.

The performance of this system is outlined in the graph of spark voltage against rpm for the actual requirement (very similar to the conventional ignition system output), and for the capacitor-discharge system.

1.5 Electronic Ignition Circuit

The circuit is illustrated in figure 1.5 and figure 1.6 since the converter circuit can be used for the generation of 240 V ac from a 12 V car battery, and for the generation of 300 to 400 V dc for other uses. This unit, as with many others in this book, does not seriously interfere with the existing car wiring and a means is provided for conversion from the original wiring to the new electronic system by merely plugging a three-pin plug into socket 1 or socket 2 in figure 1.5. The more ambitious constructor can install a change-over relay inside the unit which can be operated from inside the car as the car is moving along, for comparison between the two systems or to revert to the old system should a fault occur.

The components of figure 1.5 can be identified with those of figure 1.4, namely the 0.47 μF capacitor, thyristor and 25 Ω resistor. The bounce suppression circuit comprises a 8.2 kΩ resistor and 0.22 μF capacitor which, with a trigger diode and 1 kΩ bleed resistor, form the thyristor gating components. The only additional component is the neon, which can be mounted inside the car for

12 / Projects for the Car and Garage

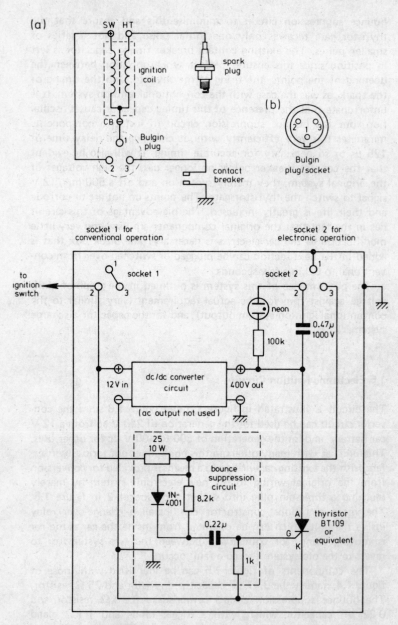

Figure 1.5 Modified ignition system. (a) Circuit diagram. (b) Bulgin plug/socket.

indication that the system is fully operational; this is of course optional and possibly unnecessary since malfunction will cause the engine to stop! The unit should be housed in a dirt-proof box near to the ignition coil, the box being well earthed and easily accessible. The thyristor and 25 Ω resistor should be mounted for good heat dissipation (away from other components or the case).

The output EHT should be well insulated for safety. The method of doing this is to use cable with a relatively thick insulation, the output voltage being several hundred volts which may arc across a thin conductor to earth and cause excess heating or complete breakdown of the conducting cable. Normal mains cable will do and the precautions normally taken with mains wiring will suffice. The wiring to socket 2 pins 2 and 3 is particularly important. Use thick insulated cable for this wiring and to the 0.47 μF capacitor at the converter output.

1.5.1 Construction Sequence

The construction of the capacitor-ignition circuit involves the circuit of figure 1.5 together with the circuit of the dc/dc converter of figure 1.2. The sequence of operations is as follows.

(1) **Construct and test the converter** circuit described in the following sections by connecting to a 12 V car battery and confirming that it delivers the necessary 300 V output.

(2) **Build the box** outlined in section 1.4 together with the thyristor board of figure 1.6 and the change-over sockets as shown.

(3) **Test the completed unit** by connecting it to a car battery and — using the connections to socket 2 (CD ignition socket) — substitute the contact breaker points by two short pieces of wire, seeing if there is a 300 V pulse produced at the output pin 2. This can be detected by connecting a voltmeter, set to read a few hundred volts, to the output pin and looking for a reading of 100 or 200 V (as a quick flick of the meter needle) as the substitute contact breaker is opened and closed. The value seen will depend on the meter damping. Remember that 300 V can be quite nasty! If nothing happens but the converter does operate correctly, then suspect the thyristor or its wiring.

(4) **Place the entire circuit into the car** as described above and test plug 1 in socket 1 first to check the conventional system wiring (anyone can make mistakes) and then finally into socket 2. Section 1.6 lists a comprehensive test procedure. A very convenient plug for plug 1 and corresponding sockets 1 and 2 is the three-pin 'Bulgin' type often used for mains plugs and sockets. These are about 20 mm

14 / Projects for the Car and Garage

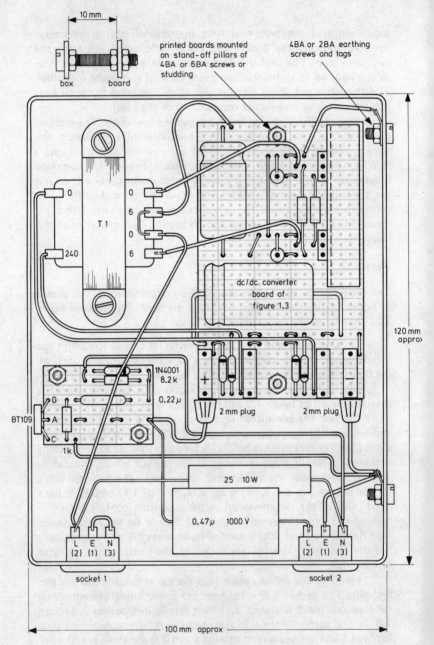

Figure 1.6   Modified ignition.

Ignition Circuits / 15

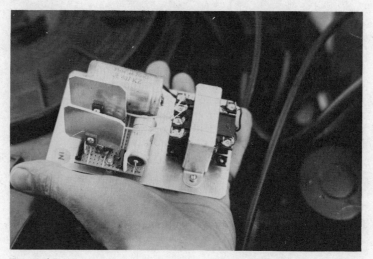

The dc/dc converter. On the prototype, I used separate heat-dissipating fins on the transistors.

round with a locating notch, the pin 1 being slightly larger than pins 2 and 3 (see figure 1.5b).

A ready-made printed circuit board for the d.c. to d.c. converter and for the complete ignition circuit is available from Maplin Electronic Supplies Ltd — see p. 108.

1.6 CD Ignition Installation

When installing the unit, first connect the power leads and listen for a high-frequency whistle from the transformer of the converter circuit. The ac and dc output of the converter circuit can then be checked with a meter. Failure to oscillate will require normal fault-finding procedures with possible wrong transformer, or BD711, connections. Beware of high voltages; this unit generates lethal voltages.

After initial checking, the ignition can be rewired as shown; first use socket 1 to check your rewiring and then use socket 2. Failure to work will again require normal checks to the thyristor and wiring, but again beware of the high voltages generated. Do **not** touch or change the EHT wiring of the ignition, this will work as normal. In the prototype the 25 Ω resistor was constructed from two 47 Ω 5 W resistors in parallel. This often provides a more reliable method of mounting power resistors since it distributes the heat generated

16 / Projects for the Car and Garage

to two, or four or more resistors; four 100 Ω 3 W resistors would be even better than two 47 Ω 5 W.

The leads to the unit should be as short as possible, using normal mains cable (the colour coding is left to you) protected against the oil and dirt of the engine and kept away from any moving parts by using straps. The final test is to try out the unit on a short run at all speeds, and check — with the engine switched off — to see that no component is overheating; additional heat sinking may be necessary (although this was not needed on the prototype).

1.7 Ignition Timing Light

The dc/dc converter of figure 1.2 can be used directly with a **xenon flash tube** (type ED70 as shown in figure 1.7 or the flash tube from

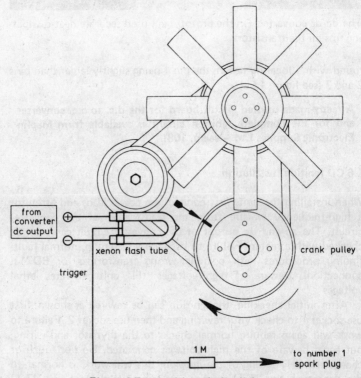

Figure 1.7   Ignition timing light.

a photographic electronic flash unit) to check the engine timing. The converter provides the 400 V necessary for the xenon tube and the trigger voltage of the tube is derived from a connection to the top cap of plug 1 via a 1 MΩ resistor. The xenon tube may be housed in a plastic tube for the safety both of the handler (who would be surprised by 400 V dc or 40 kV EHT from electronic ignition), and of the tube, whose connections are very fragile and liable to break off. A tube of 20 mm inside diameter should be adequate, the tube and 1 MΩ resistor being glued in place. The car is then moved into a dark place, and the unit connected up and placed near to the engine front where the crank pulley and crank case markings are situated. Remember to keep all wires away from the moving parts (fan belt, fan, and so on) when doing this. Start the engine, and the tube will now flash in time with the engine; stroboscopic action will allow the timing to be adjusted, using the distributor micrometer vacuum screw to advance or retard the ignition to that recommended by the manufacturer. In some cars the whole distributor housing is rotated until the marks are correctly aligned — consult the manual or your local garage for full instructions.

1.8 A Dwell Meter

One adjustment that is difficult to make is the setting of the contact breaker gap with feeler gauges which should slip between the points when the gap is fully open. This involves pushing the car along in gear with the uncertainty of not knowing that the gap is set correctly; the feeler gauges often push against the spring and give a false setting. The **dwell meter** allows the gap to be measured while the engine is turning, and figure 1.8 shows what is meant by the **dwell angle**; this is the time, measured in degrees of rotation, that the contacts are **closed**. The circuit senses this angle by measuring the time that the contact breaker points are closed using the given circuit, the angle being read on a 1 mA meter duly calibrated in degrees.

The signal appearing across the points is seen in figure 1.8 to be a form of damped oscillation; this is converted via the bridge rectifier and 5.1 V Zener to a squarewave of 2.5 V at TR1 base. The signal will be positive as the contacts open and zero when closed. When open, TR1 is turned on to short-circuit TR2 base—emitter and prevent any current flowing in the meter. During the dwell period TR1 is turned off to allow TR2 to conduct and feed current to the meter during this period, in one direction only, from positive to negative.

18 / Projects for the Car and Garage

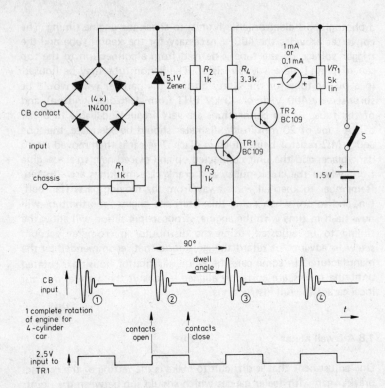

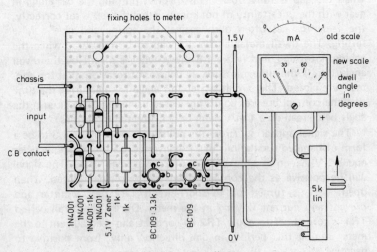

Figure 1.8 Dwell meter.

As the points open and close, therefore, the meter is fed with a series of positive impulses which will average out to give a direct reading of angle, $0°$ to $90°$. A six-cylinder car will have a dwell angle of less than the $40°$ to $60°$ of a four-cylinder car since the total period between pulses is only $60°$; consult the manual for the recommended angle. The components can be mounted on a single printed board mounted with the meter and variable resistor in a suitable box. The polarity of the input leads is unimportant, but the meter may need to be reversed for positive readings only. VR_1 is adjusted for a $90°$ reading on a shorted input; the new scale can then use the same scale numbering as 0 to 1 mA or 0 to 100 μA.

To use this meter

(1) Calibrate the meter as above and connect to chassis and the CB on the distributor of the car.

(2) Remove the distributor cap and rotor arm and loosen the contact breaker point screw for adjustment purposes.

(3) Remove all the spark-plugs from the engine but mark the plug leads accordingly.

(4) Switch on the ignition and use the starter motor to turn over the car, noting the dwell angle reading.

(5) Adjust the point's gap by inserting a screwdriver into the slot of the adjustment screw (see figure 1.1).

(6) Tighten the screw, replace the cap plugs and the leads.

(7) Carry out the tuning adjustments using the xenon tube as described earlier; dwell time should be followed by the timing adjustment.

1.9 Ignition Suppression

Car radio interference can be caused by many aspects of ignition interference. Some are obvious but many are difficult, or even impossible, to trace. The interference is generated by all forms of arcing in the ignition circuit or by contact breaking in all switches and motors including indicator flashers, solenoids, relays, windscreen wipers, brake light switches, and so on. The main source is in the ignition circuit and the steps that can be taken to minimise interference are as follows.

(1) Use screened plug caps (available from suppliers).
(2) Use carbon-string EHT leads.
(3) Install suppression caps to the distributor top or in-line suppressors in the plug leads.

(4) Install an ignition suppressor capacitor as in figure 1.9a if not already fitted.

Dynamo suppression can be included by using one or all of the methods shown in figure 1.9, each of which forms a filter circuit to minimise the effects of commutator or brush contact noise. Check with the wiring diagram of your car to see which methods are already installed — others can be added if necessary; the capacitors that are installed may even need replacement owing to ageing or damage or the contacts may require cleaning and remaking.

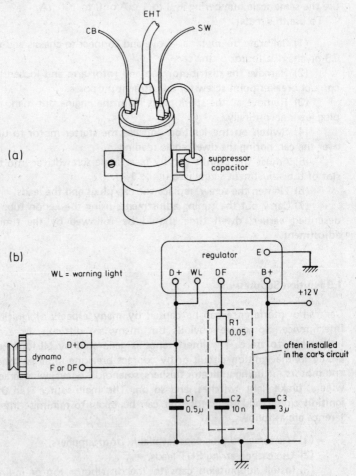

Figure 1.9 Ignition suppression. (a) The suppressor capacitor in

Noise from other motors in the car, windscreen wiper and heater fan, for example, can be suppressed by a parallel 3 μF 25 V capacitor. Electrostatic charges picked up from the road by the wheels can also cause interference but a good electrical contact formed by a spring in the wheel hub will discharge the wheels; this spring must be inserted between the wheel hub and the axle.

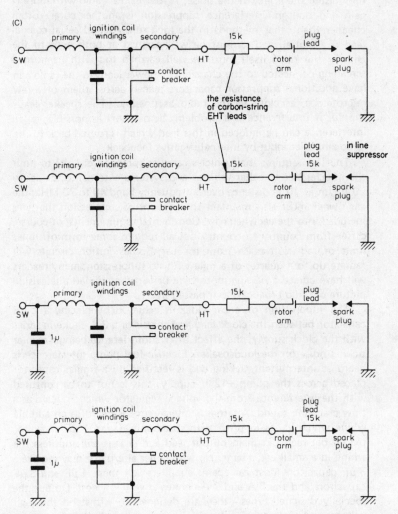

position. (b) Regulator suppression. (c) Ignition circuit suppression.

High resistance welds in the car chassis are a common source of interference and it is possible to weld shorting straps across each body section until the source is found — but this is a very expensive business. The purchase of a better car aerial or the repositioning of the aerial and radio often cures an otherwise very frustrating problem. One circuit described in chapter 5 shows an aerial amplifier that will boost the signal and improve the signal-to-noise ratio thereby minimising the effect of the noise. A 'better' car radio with more rf gain and built-in interference suppression is another cure — often cheaper than other methods in the long run. A good aerial coaxial lead is essential, and it must be well earthed at both ends to the chassis; the radio itself must be well earthed too with its normal screening connected to the chassis. The power leads to the radio can have additional suppressor capacitors placed across them of a few microfarads, as close to the radio as possible. The speaker leads should, if interference is a problem, be as short as possible; much interference can be induced in this lead which is routed back to the radio amplifier input by internal negative feedback.

The law requires that vehicles carry suppression circuits to limit the interference to other vehicles and systems external to the car to 50 μV/m at 10 m distance over a frequency band 40 to 70 MHz; the car owner must also **maintain** the amount of suppression that was included into the car when new. Continental requirements, of course, differ from country to country but all require some form of upper limit of radiation noise from the car. Some ignition circuits will radiate up to a quarter of a mile with no suppression; many readers will have cursed a passing motorbike or van that ruined a television picture or radio broadcast as it passed by.

The suppression of a car clock is achieved by placing a 1 μF capacitor between the clock feed and chassis or a 3 A choke in series with the clock supply, the effect of this interference being a regular ticking noise on the loudspeaker. Electric fuel pump interference is heard as intermittent ticking and is rectified by a similar capacitor placed across the pump +12 V supply; this is not to be confused with the interference from the voltage regulator which is heard as a slow crackling sound. Alternator interference is heard as an audible whine which varies in pitch with speed and load; a 1 μF capacitor placed between the main output lead and chassis will suppress the whine in a small car, a larger car with larger alternator may require a 3 μF capacitor. Most car spares suppliers sell these 1 μF and 3 μF capacitors and the 3 A and 7 A chokes at very low cost. Do buy the specially designed types — they are designed to withstand the high voltages and currents, oil, vibrations, water and sludge.

Glass fibre bodies present a special problem since suppression demands that the **entire** engine compartment is fully screened with tin foil or similar conducting material. The positioning of the radio receiver can minimise noise in these cars — try placing the radio at the other end to the engine together with the aerial and speakers; alternatively invest in a new **metal** car, or resign yourself to no radio or tape player.

1.10 Ignition Booster Circuit

Section 1.2, under item (5) stated that the EHT spark to the sparkplugs could be boosted in order to increase the overall efficiency of the ignition circuit. The capacitor-discharge ignition circuit achieved this by creating a 400 V supply to the ignition coil. A far simpler method of boosting the spark energy by 20 per cent or so is to increase the car battery voltage to the ignition circuit by placing a battery in series with the 12 V supply.

Figure 1.10 illustrates the method used where two nickel—cadmium batteries are switched into circuit by S1 when starting the engine. Sintered plate nickel—cadmium batteries can deliver several amps of current for short periods with no deterioration; they can also be charged up again when the engine is running normally. S1, in the 'ON' position, connects the two Ni—Cd batteries between the ignition switch (connection II) and the ignition coil, the total supply to the ignition circuit now being 14.4 V, an increase of 20 per cent. In position 'OFF', this switch returns the ignition circuit to its normal position but allows the two Ni—Cd batteries to be charged up via R_1 and R_2, two current-limiting resistors which allow about 6 mA of charging current to flow. S1 is a double pole change-over switch mounted on the car dashboard, which is operated only when starting the engine. If kept in the 'ON' position the two batteries would soon be discharged completely and the ignition circuit would gradually die.

One minor modification is to replace S1 by a small 12 V relay with similar contacts to S1 but operated when the ignition switch is turned to position II to operate the starter solenoid. In this case the relay coil is connected between point II and earth, the relay contacts being wired as shown for S1. This ensures that the Ni—Cd batteries are only in circuit during starting; the prototype was operated, however, using a switch with a small indicator light (a single LED with series 470 Ω resistor) across R_1, the right way round of course.

24 / Projects for the Car and Garage

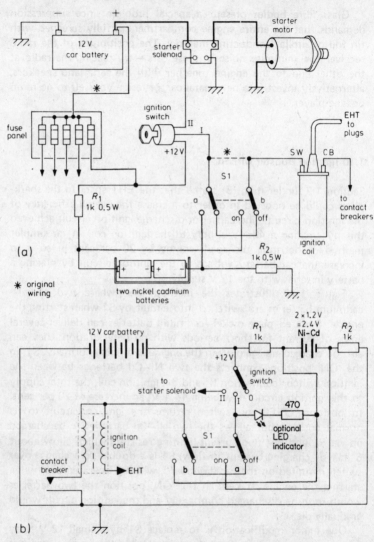

Figure 1.10 Ignition booster.

This light comes on while the Ni–Cd batteries are charging but it is short-circuited when S1b is in the 'ON' position. This serves only as a check to indicate that S1 is 'OFF' and that the two batteries are charging; the lamp is on even when the ignition switch is 'OFF' but the current drain is negligible.

2
Car Theft Circuits

There are several car theft devices available which

    (1) **Immobilise** the car preventing it from being driven away; or

    (2) **Activate an alarm** circuit if an intruder does not give the magic password or start the car in the correct sequence — these are deterrent devices; or

    (3) Activate the alarm if the car is **touched** (another deterrent).

This chapter describes simple versions of each. None of these methods interferes with the existing wiring of the car, they merely involve placing additional circuits/devices across the existing switches or lights or horn.

2.1 Immobilising Circuits

Figure 2.1 shows three such circuits where

    (1) The 12 V feed to the ignition coil (SW terminal) is open-circuited (S1); or

    (2) The contact breaker points are permanently short-circuited (S2); or

    (3) the +12 V feed to the starter motor solenoid is open-circuited (S3).

These switches may be placed inside the car at some convenient point and the wiring concealed. Switches S1 and S3 may be placed very near to the ignition switch whereas S2 involves one lead from the distributor or ignition coil to the car dashboard. It should be pointed out that the placing of series components such as S1 and S3 into the car wiring does marginally reduce the reliability of the circuitry, adding additional components which themselves can go faulty. Good reliable switches and quality wiring should therefore be used.

2.2 Alarm Circuits

Switches S4 and S5 of figure 2.1 illustrate two alarm circuits where

(1) The horn is sounded automatically every time the ignition switch is operated. This is for cars with one side of the horn earthed to chassis — S4 disconnects this device.

(2) The horn again sounds with the starter motor but this is intended for cars where one side of the horn goes to +12 V. A relay completes the circuit, disconnected by operating S5.

The sound of the horn blasting away should act as an active deterrent but you may wish to prolong the agony with a **latching relay** as shown in figure 2.2. The relay has two sets of contacts, one used for operating the horn direct, the other for latching the relay. This device can only then be disconnected by pulling the +12 V lead marked 'X' off the battery, or off the ignition switch as appropriate, or a secret switch can be hidden inside or outside the car to immobilise the alarm.

One further addition to this latch is a circuit which sounds the horn for 1 min or for 5 min by using one of two circuits to be found in chapter 3, namely a 'nightlight' circuit which switches +12 V to the alarm circuit at point A for 5 min or the 'interior light extender' circuit which performs a similar job but only for 1 min. This will prevent the horn from draining the battery if the occasion arises.

The **mercury switch** is a very popular means of operating any of these latch alarms as the car is rocked — primarily intended for operation by an intruder but very often they are set off by passing traffic or the wind. The mercury switch, a very low cost device, is secured at an angle somewhere on the car chassis along with the time-delay and relay circuits hidden out of view so that the thief cannot see how to immobilise the alarm quickly. A position behind the dashboard is possible since there is access to the ignition switch, horn button, chassis and +12 V at this point. The connection to

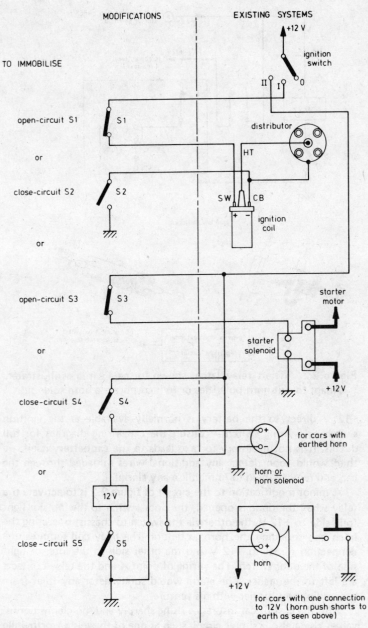

Figure 2.1 Theft circuits. The switches are shown in normal position.

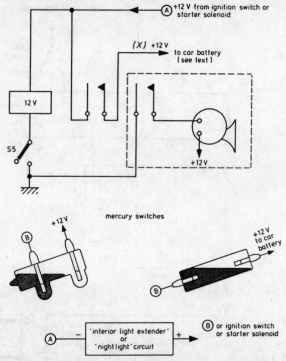

Figure 2.2   Theft relay latch, shown for cars wired as illustrated. Adapt to suit horn polarities or to incorporate a horn solenoid.

+12 V direct to the battery is normally available at the ignition switch or lighting switch — consult the car wiring diagram for full details. It is not necessary to take leads to the car battery itself — a thief would soon detect any additional wires threaded through the car, and remove them to immobilise any alarm!

A minor modification to the circuit of figure 2.2 is to activate the relay when the door is opened, the door switch to the interior light (one side to +12 V, the other via the switch to chassis) operating the latch and sounding the horn as before. The relay coil requires one connection direct to +12 V and the other side to the interior light side of the door switch. The wiring of point X and the relay contacts are left to the reader. The alarm would surely deter any thief from proceeding any further with his venture.

With the cost of car insurance rising sharply and 'no-claims bonus' values rising also, a simple circuit such as one of these is a worthwhile investment.

3
Lighting Circuits

One section of the car electrical system which lends itself to low voltage electronics is the **lighting** system. Now that high current transistors such as the 2N3055 are readily available at very low cost, there is no problem in switching 4 A to the headlamps, 2 A to the direction indicators or stop lamps and 0.5 A to the small indicator lamps — the 2N3055 being capable of switching up to 8 A or more for short periods. This does not mean, however, that this chapter should be renamed '101 uses of the 2N3055'; in fact the 2N3055 is **not** used at all — the BD131/2 is preferred since it is smaller and does not need the heavy base current of the 2N3055. The BD131 and BD132 (*npn* and *pnp*) are rated at 6 A continuously (or more if pulsed), and so are adequate for most uses. A more powerful plastic transistor is the 75 W BD711 which can cope with 12 A, the size and connections being the same as the BD131.

The lighting system of a car uses one side of most lights as chassis (0 V), and the switches activate 12 V to make the complete circuit. One light which is often used in this book is the **interior light** and its associated door switch(es) and the reader is urged to ensure that these are in order before connecting wires in parallel. In this way any alarm(s) will not sound while the car is moving since this might endanger the driver and passengers; this moment of excitement is delayed until he opens his door to leave the car.

3.1 Interior Light Extender

The circuit of figure 3.1 places a time-delayed open circuit across the door switch(es) to slowly reduce the brightness of the interior light after the car door is shut, thereby allowing the driver to find his keys, the ignition lock and seat belt in about 15 to 20 s. The circuit operates when the car door is closed and the door switch goes from closed to open.

With the door open the 470 µF capacitor is fully discharged but as the door closes (switch opens) this capacitor slowly charges up via the 1.5 kΩ resistor. In the meantime — while charging takes place — the BC177 is turned on, which turns on the BD131 and the lamp stays alight. Gradually the capacitor charges to turn off the BC177 and the BD131 until after 15 to 20 s both are fully off and negligible current flows. The 1N4001 diode allows quick discharge, when the door is opened and shut quickly, to activate the circuit again. Although the interior light of most cars is only 6 W, some cars may require lights with currents of a few amps to be lit and a small aluminium heat sink should be bolted to the BD131 as shown in figure 3.1. The dimensions shown are not critical and it is helpful to remember that the metal side of the BD131 is connected internally to the collector, so watch for short-circuits if the metal chassis is used as a sink. This basic circuit has other uses in chapter 2 and later in this chapter; it may be placed anywhere in series with the supply provided the correct polarities are observed.

The interior light extender circuit is very compact — but the heat sink, insulated from all other components except the transistor, is essential.

Lighting Circuits / 31

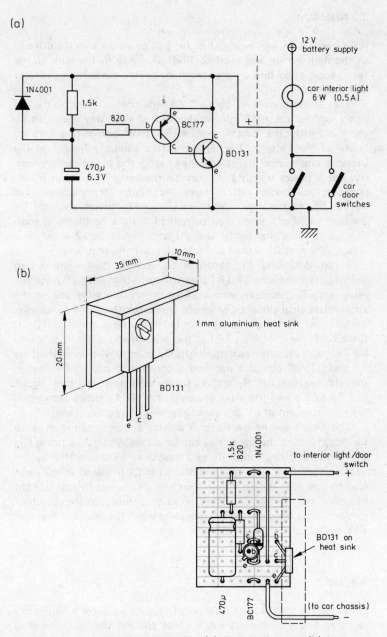

Figure 3.1 Interior light extender. (a) Circuit diagram. (b) Layout.

3.2 Nightlight

Have you ever driven home at night, parked your car in the drive or in the dark garage and tried to find your way in the dark to the front door, or to find a light switch, or to find the correct door key and lock?

The circuit shown in figure 3.2 keeps one or more of the car lights 'on' for up to 10 min while you find your way about, lit by the car lights. The circuit then switches off completely with no waste of the battery. The circuit uses a similar principle to the previous one except that it activates a relay and so avoids the necessity for a power transistor to turn on the lamp. The circuit is not, however, automatic and requires the momentary pushing of a button S1, to start the time-delay. The relay chosen should have contacts rated at the required current (4 A for a headlamp or spotlight; 2 A for a side light), and the coil should be 12 V or thereabouts (the prototype used an 18 V relay with no problem).

Before operating S1, capacitor C_1 is fully discharged and so initially after pressing S1 TR1 base voltage is high with TR2 and the relay 'on'. The contacts across S1 latch 'on' the relay and so the appropriate sidelight(s) or headlight comes on. Gradually C_1 charges up to lower TR1 base voltage until, after several minutes, TR1 is turned off (the voltage on TR1 at this point is about 2.5 V) followed by TR2 and the relay switching off. C_1 quickly discharges itself via R_1 and D1; D2 acts as a transient suppressor to protect TR2 when the relay switches off. R_2 and R_4 form a 'bootstrap' feedback circuit in TR1 to increase the input impedance of TR1 thereby prolonging the time constant of C_1 and generating very long delay times.

This circuit can be used also as a delay circuit to sound an alarm for 5 to 10 min; the time-delay can be varied by merely altering C_1, a larger value giving longer delay times; since S1 is operated by the ignition switch or a mercury switch, or if S1 is placed in the earth lead by the door switch, this circuit will also act as an interior light extender, the connections being wired accordingly. 'Normally open' contacts on a relay or microswitch **close** when the device is operated whereas 'normally closed' contacts **open**.

3.3 'Lights-on' Reminder

How many times have you returned to your car after a hard day's work to find your battery flat because you left the lights switched on all day? The circuit shown in figure 3.3 will sound an alarm if the

Lighting Circuits / 33

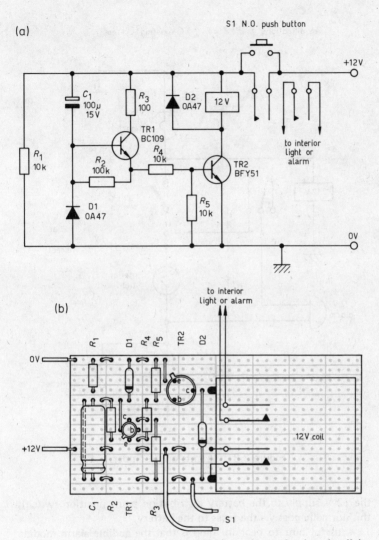

Figure 3.2   Nightlight (or theft delay) or alternative interior light extender. (a) Circuit diagram. (b) Layout.

lights are on and you open the door — the alarm used being either the small **audible alarm module** available at low cost from 'Maplin' or the circuit of section 3.4. With both doors closed the relay will be activated every time the lights are switched on; the alarm only sounds, however, when the door is opened. Remember to connect

34 / Projects for the Car and Garage

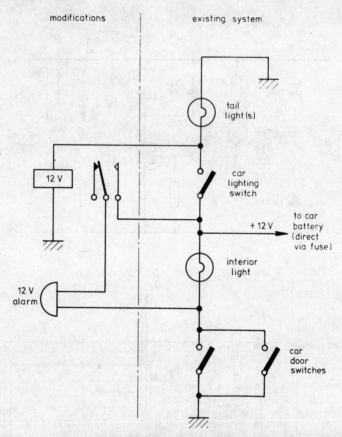

Figure 3.3 'Lights-on' reminder.

the 12 V supply to the battery distribution point (ignition switch); this normally goes via the fuses to the battery.

A useful hint to bear in mind is that the **audible alarm module** volume may be increased in any one of these circuits by placing the alarm in series with an ordinary loudspeaker whose output will be clearly heard. The module consumes about 80 mA of current.

3.4 Alarm Circuit

An alternative to the audible alarm module is the circuit of figure 3.4 using a 555 timer IC and any loudspeaker; the total cost is about

Lighting Circuits / 35

The circuit can be made on a very small piece of Veroboard

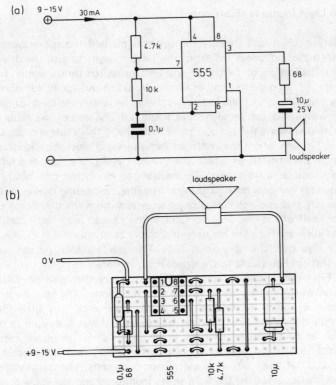

Figure 3.4 Alarm circuit. (a) Circuit diagram. (b) Layout.

the same but volume output is greater and current is less, about 30 mA. The tone may also be changed by varying the 0.1 μF capacitor or 10 kΩ resistor.

3.5 'Ignition Key In' Reminder

After once having had to call the AA to break into my car and retrieve my car keys from inside, I should then have installed the simple device shown in figure 3.5, which will also sound an alarm if the key remains in the lock and the door is opened. The alarm is the same as before, the microswitch being mounted in any convenient place near the ignition switch on the steering column or on the dashboard, according to the car.

3.6 Light Monitors (Fibre-optic)

Many modern cars (mainly the overseas models) incorporate small fibre-optic indicators fed from the lighting units to give the driver some indication of failure of any light or switch (brake switch for example). With the low cost of fibre-optic cables and epoxy adhesives, the installation of these indicators should be within the reach of any car owner. As can be seen from figure 3.6, one end of the cable is pointed towards the lamp(s) to be tested, and this cable end can be in a housing where the left/right indicators and sidelight and headlamps are situated or where the tail light and brake light and left/right indicators are housed. Connections to more than one lamp (if required) are now possible using plastic interconnecting sleeves, only one left and one right cable being necessary from the dashboard or rear shelf of the car. The indicator in the car can use a small plastic lens such as that used on neon indicators or on calculator displays; this slips over the fibre-optic cable. The two indicators on the rear shelf should be visible to the driver in his rear-view mirror.

Fibre-optic cable comprises a bundle of very thin glass or plastic fibres which are contained within a plastic sheath. The light passing into one end is **transmitted** to the other end with very little loss within the cable; the stranded construction allows the cable to be bent in all directions without breaking. In order to install these cables into a car as light monitors, the cables must be prepared correctly at their receiving and transmitting ends. The transmitting end is placed near to the light being monitored and the cable must be cut cleanly across the fibres and then sanded down with fine

Lighting Circuits / 37

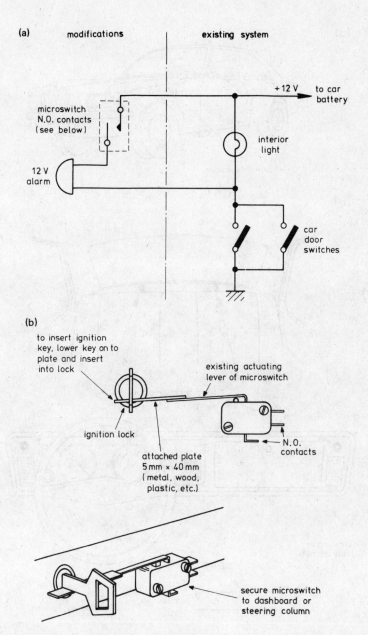

Figure 3.5 'Ignition key in' reminder. (a) Circuit diagram. (b) Layout.

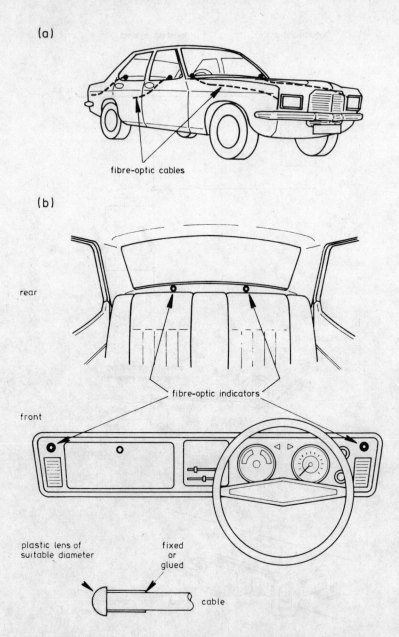

Figure 3.6 Light monitors. (a) Fibre-optic cables. (b) Fibre-optic indicators.

Lighting Circuits / 39

emery paper until very smooth. The end can then be polished with buffing solution if desired. The receiving end is prepared in a similar way so that no light is reflected back down the cable (this would reduce the light output). The plastic lens merely increases the light area so that it can be seen from a distance — on the rear shelf of the car, say. The lens should not be glued on to the polished smooth surface of the cable since this will disperse the light; it should be fixed or glued to the edge of the cable, as shown.

Many fibre-optic suppliers offer kits which contain all the necessary polishing ingredients, lenses and instructions; a little experimenting with these cables will inevitably result in the best positioning of transmitting and receiving ends.

3.7 Electrical Car Light Failure Monitor

Figure 3.7 shows a system which monitors the failure of **one** light in a pair: sidelights, or headlights, or tail lights, or stoplights. Counter-wound coils are wrapped around a small reed relay so that the magnetic fields created by the currents to both lamps cancel and the relay does not operate. If one lamp fails then the reed relay switches on a warning lamp, or LED in series with a 470 Ω resistor, to 12 V. Four or more such relays will be required for all the pairs of lights, and these must be positioned near to the lights at the point where the current feed splits to the left and right lights.

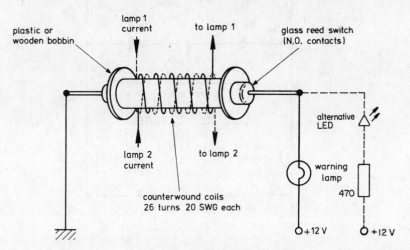

Figure 3.7 Electrical car lights failure monitor.

3.8 Audible Turning Indicator

Are you hard of hearing? ('Pardon?') Can you hear the click of the flasher unit as it operates in your car? In many modern cars this unit is far away from the dashboard where it cannot be heard and often the flashers wink away unknown to the driver, thereby creating havoc on the roads filled with frustrated drivers not knowing when or where the car is going to turn to left or right. Figure 3.8 shows a circuit which sounds an alarm every time the indicator arm is operated, and very little additional wiring is necessary. The alarm devices already described are used, these being positioned near to the steering column where they may be heard and near to the indicator switch.

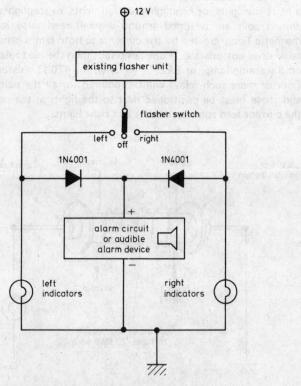

Figure 3.8   Audible turning indicators.

3.9 Emergency Flasher

This circuit (see figure 3.9) can be added to existing flasher units to enable **all** left/right indicator lights to flash together. The circuit is simplicity itself, with the existing flasher unit acting as the master and the additional relay as slave to operate the left indicators; the 12 V relay contacts should be rated at 2 A for most cars, but please confirm this by consulting the handbook. The existing flasher will not notice the addition of the relay via S2a, most relays taking only

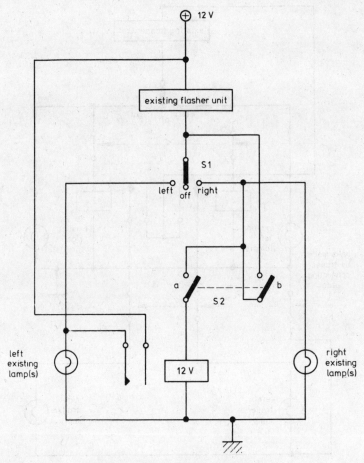

Figure 3.9 Emergency flasher.

10 to 20 mA; S2b merely bypasses the normal indicator switch S1 which will operate in the usual manner.

3.10 Trailer Flashers

The parallel connection of trailer flashers would normally drastically alter the flashing rate of the flasher unit and possibly overload the

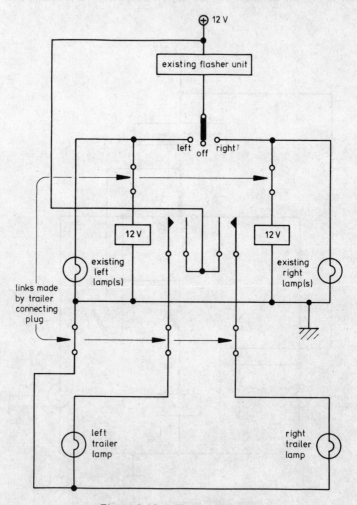

Figure 3.10   Trailer flashers.

unit as well. The addition of two 12 V relays, as in figure 3.10, overcomes this problem and the wiring is very simple indeed. The connecting plug — an octal plug and socket — can be used to make the five links necessary, the three remaining connections being used for stop lights, tail lights and number plate indicator (the earth connection takes up one of the five already used).

4
Accessory Circuits

The circuits described in this chapter are for the motorist who likes to drive in safe comfort, with pleasing 'noiseless' sound from his car radio, a crystal clear windscreen in all weathers, the correct time and a clear indication of the condition of his battery and engine. He also wants to have an emergency beacon to stand by his car to warn passing motorists, if an emergency does arise. Many of these devices are standard fitments in modern cars (mainly overseas models) but for those, like me, who have yet to afford such pleasures, electronic 'know-how' is the solution. Most of these circuits should be separately fused from the 12 V battery and all assume a negative chassis connection with a switch in the positive lead. It may be worth while fitting a dashboard panel into the car which incorporates the circuits of this chapter — some motorist supermarkets sell ready-cut boards with holes for meters and switches; the small circuits can be mounted on the rear, all fed from a common fused 12 V supply.

4.1 Car Radio Audio Booster

One major difference between an expensive and a low-priced car radio lies in the audio output power and hence in the quality of reproduction, particularly at bass frequencies; the radio frequency circuits of the two are very similar. It is possible to add a more powerful output stage to a low-priced radio or to a cassette player with just a few components.

Many integrated circuits are available on the market that will provide several watts of audio power for a car radio or cassette player. A few of these are listed in table 4.1.

Table 4.1

Pre/main amps	Power rms (W)	Input sensitivity (mV)	Speaker (Ω)	Supply (V)	Price each (for comparison only)
LM379	2 × 6	120	8	10–35	£6
LM380	2	80	8	20	£1.27
LM381	–	–	–	9–40	£2.13
SL414A	3	250	8	20	£2.86
SN76003N	5	28	15	35 Max	£3.22
SN76013N	5	20	8	28 Max	£2.45
SN76023N	5	–	8	28 Max	£2.45
SN76033N	5	–	15	35 Max	£3.22
TBA800	5	80	16	30	£1.25
TBA810S	7	35	4	20	£1.40
TBA820	2	16	8	16	£1.00
TDA2002A (or 2003)	7	55	4	5–25	£1.60
TDA2020	20	80	4	30	£3.20
TDA2030	4	10	4	12	£3.20

A glance at these circuits shows that the LM380 and TBA820 are the best 2 W amplifiers for 8 Ω output with the TBA810S the best high wattage amplifier for 4 Ω output. All are suitable for 12 V operation and a selection of circuits follows, each of which can be connected to the car radio or cassette player at the point shown in figure 4.1.

Comparisons between the circuits shown in figures 4.1, 4.2 and 4.3 show that the LM380 provides the 'simplest' circuit, including volume; an additional **tone control** of series 10 kΩ linear pot. and 0.05 μF capacitor can be placed between pins 2 and 6. The TBA circuits require many coupling and decoupling capacitors which increases cost and size. The TDA 2002 is a purpose-made car radio output IC with transient pulse protection but the cost of this IC is higher. Finally the LM379, a double 8 W IC provides 16 W of power

46 / Projects for the Car and Garage

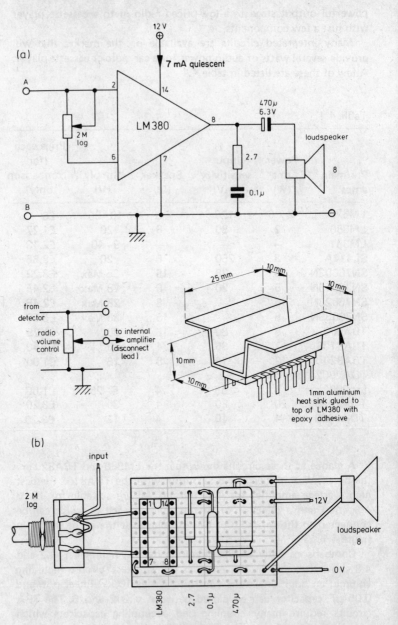

Figure 4.1 Car radio audio boosters: 2 W (six components, including volume). (a) Circuit diagram. (b) Layout.

Accessory Circuits / 47

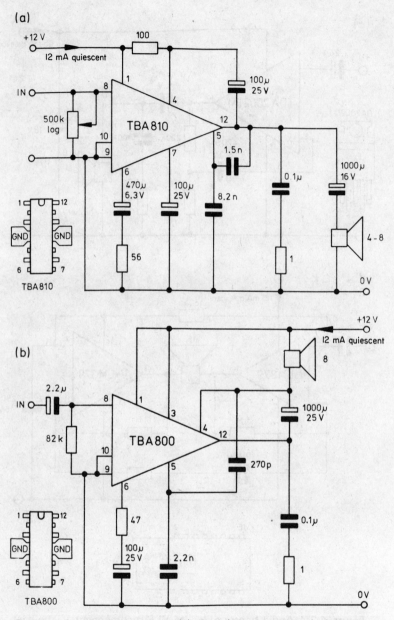

Figure 4.2 Audio boosters. (a) 5 W (13 components). (b) 4 W (11 components).

48 / Projects for the Car and Garage

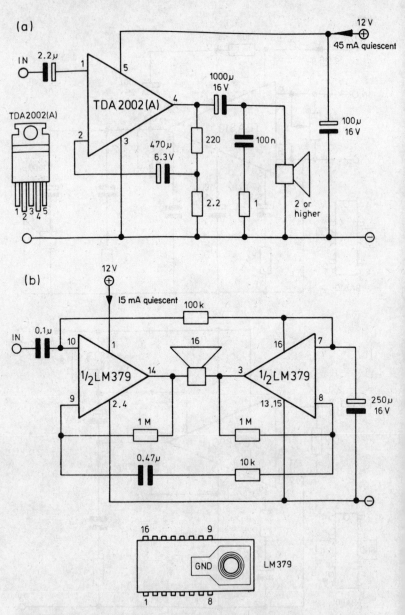

Figure 4.3 Audio boosters. (a) ≥4 W (10 components). The heat sink requires 45 cm² of 3 mm aluminium. (b) 16 W (nine components).

with a minimum of components which should be enough for the deafest of motorists. All these circuits should

(1) Have their input power leads **decoupled** with a 470 µF, 15 V capacitor.

(2) Be placed as **close** to the radio or cassette player as possible.

(3) Be connected at point C in figure 4.1 if the circuit contains its own volume control or point D if the radio volume control is to be used. However, disconnect the capacitor which leaves the wiper of the volume control to go to the radio amplifier or the differences in dc level may cause problems.

(4) Be adequately **heat sinked** with a flow of air available to cool the IC. Do not mount the IC or heat sink in a position where the full blaze of the sun can cause overheating. In many cases the heat sink provided can be bolted to the car chassis thereby providing a ready-made conduction heat sink.

(5) Have the external components mounted as close to the IC as possible so that **high frequency instability** does not cause problems such as excessive background noise.

The series resistor and capacitor across the speaker are intended for hf instability prevention. Figure 4.4 shows a push—pull version with two LM380 ICs to deliver 4 W of power into a 16 Ω speaker. Many of these may be connected in this manner; the 1 MΩ preset balances the two amplifiers; this is set up by placing a 500 mA meter in series with the speaker and setting the preset for minimum current, under no signal conditions.

The TDA2020 and TDA2030 can operate on single 12 V supplies — see *Electronic Projects 3* for circuits.

Finally, watch the **power rating** of the speaker in watts, obtained from the manufacturer's literature supplied with the car radio. This may be under-rated for the booster amplifier and smoke may rise from the speaker coil under full power!

4.1.1 Stereo Systems

The majority of tape players on the market these days are designed for stereo operation, with the left and right speakers inserted into the doors, door frames or the rear compartments. All the amplifier circuits described in section 4.1 can be used for stereo operation; many have built-in balance components which, after the connection of an external balance potentiometer, allow two ICs to be interconnected for this purpose.

Stereo radio presents some problems. The receiver must operate on VHF and incorporate the usual fm multiplex decoder which

50 / Projects for the Car and Garage

(a)

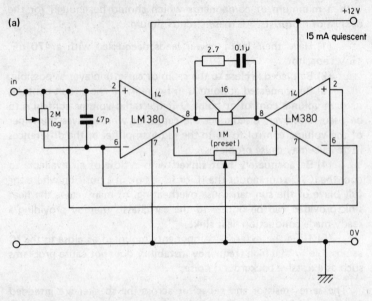

(b)

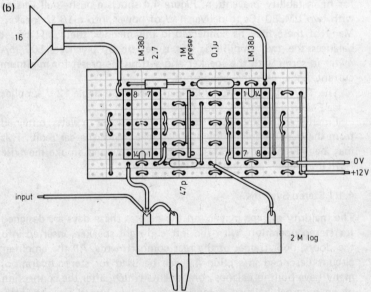

Figure 4.4 Push–pull audio amplifier; 4 W (eight components). (a) Circuit diagram. (b) Layout. The LM380s must be mounted on heat sinks as in figure 4.1

struggles to pick up the stereo signal from the car aerial. Chapter 1 outlined the suppression problems of modern cars and the attempts made to minimise this interference at the source and in the radio receiver or amplifier. Most of this interference is amplitude modulated and so will automatically be rejected by the fm receiver circuits — this is one point in favour of the fm car radio. The point against fm car radios and, in particular, stereo car radios, lies not in the car but outside. The fm radio must pick up a relatively high amount of signal on the VHF band, far more than the am radio signal. A normal domestic fm radio uses a split dipole aerial of length 1.5 m, this being half of the 100 MHz wavelength corresponding to 3 m ÷ 2, and it must point in the right direction for good signal strength. Added to this is the fact that a VHF frequency modulated signal travels only a limited distance and is easily absorbed by objects in the signal path, including fog and rain, and also the fact that the signal can get distorted by reflected signals which add to or subtract from the direct signal to produce a phasing effect. This can be heard on a fm radio as an aircraft flies overhead when the characteristic flutter effect is produced as the aircraft puts the reflected fm radio signal alternately in and out of phase as it flies by.

Imagine a car moving through streets with an aerial placed on top, with the car going up and down in all directions. This is just not the way to get hi-fi sound! One solution is to incorporate a very high gain amplifier at the aerial input with a large amount of automatic gain control (agc) to compensate for the varying signal amplitude; this would not compensate for the phasing effect and so good stereo reception is almost impossible. When digital stereo transmissions begin the phasing, amplitude and noise effects will be unimportant but until that day, keep the stereo for tape players and use normal am radio circuits which — with present technology — can sound almost as good as fm. And if stereo sounds are a must, then use a pseudo-stereo system as outlined in the next section.

4.1.2 Pseudo-stereo Systems

There are several ways of confusing a listener into thinking that the sounds from two loudspeakers are true stereo. One method is to separate the frequencies into bass and treble, routing each group of frequencies to one speaker. Another is to phase the mono signal differently to two 'stereo' speakers, and a third method allows the mono signal to slowly drift from one speaker to another on a rotation basis. The second method is used by recording companies to produce pseudo-stereo and it produces the echo-bathroom effect often heard on old recordings that have been processed; the third method requires considerable additional circuitry to rotate the volumes of the signals around and so I have settled for the simplest solution, which is the filter method.

The filter used can be of the active type shown in figure 4.5b or the passive type shown in figure 4.5a. Both will produce the **crossover frequency response** seen in the diagram with the treble speaker starting to cut off at 400 Hz, the bass speaker starting to cut off at 2 kHz. Three speakers are shown in the passive circuit: one may be placed in the front or back of the car, the others can be placed in the doors; the middle frequency speaker can be omitted if two are found to be enough. Remember that the amplifier output will be split between all three speakers and so the power, 4 W or so, will be split into about 1 W per speaker, with some being wasted in the frequency-splitting network. The inductors used can be purchased from component suppliers, as can the capacitors, but these must be capable of passing the large output currents to the speakers; the capacitors should be **non-polarised** types such as paper, polycarbonate, or polyester (normal electrolytics will **not** do).

The types of active filter are numerous, but since this application is merely to produce an approximate pseudo-stereo, the simple circuits shown can be used, each circuit having a 6 dB per octave fall at the given frequencies — enough to split the bass and treble in the car. Remember that these sounds will be superimposed on top of the engine and wind noise as the car drives along so do **not** expect high Hi-Fi in your car unless you have a Daimler or Rolls (in which case you won't be installing substitute stereo — you will already have the real thing). The power amplifiers used can be any of those listed in this chapter or **one** can be the internal radio or tape amplifier, the **other** being an additional booster amplifier, such as one of the IC types; do not connect these two amplifiers to the output speaker terminals of your radio or tape player — smoke will rise from within! Connect them to the **input** of the internal amplifier or the output of the internal pre-amplifier — this is usually at the point where the volume control is situated. If this is the case, there is no reason why you should not use the now-obsolete internal amplifier as one of the power amplifiers since, as we are not interested in perfect balanced stereo here, we only want to split bass and treble.

Assuming you now have stereo operation, whether true or pseudo, you should now think **quadrophonic**. There are quad systems available which run from four or eight-track tape machines and these can be installed into a car with ease. Quad radio is, however, even more difficult to operate than stereo and so pseudo-quad is the only solution. Two such circuits are shown in figure 4.6: one very simple

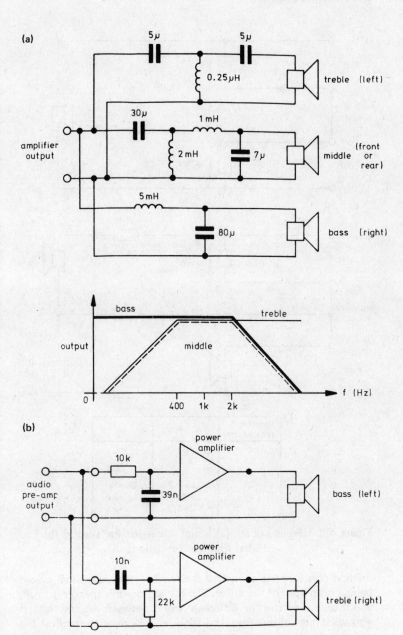

Figure 4.5 Pseudo-stereo. (a) Passive cross-over unit used for pseudo-stereo. (b) Active unit.

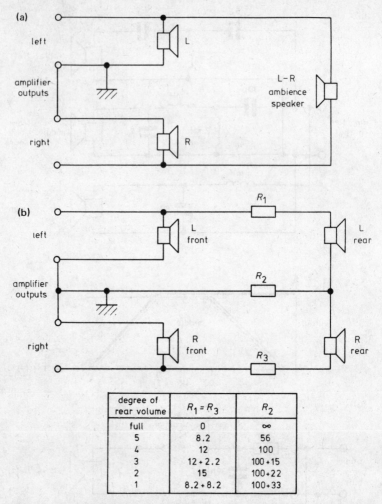

Figure 4.6 Pseudo-quad. (a) Simple three-speaker system. (b) The Hafler four-speaker system.

method uses just **one** additional speaker for **ambience**, either placed behind or in front of the driver. The sound from this speaker is small, and it only produces the difference sound between left and right; if you want the volume from this speaker to increase then adjust the volumes from all three speakers either by placing a few ohms in series with the main left and right speakers or by mismatching the main speakers. For example with a 8 Ω system, the ambience speaker can be 8 Ω, but the left and right speakers can be 12 or 15 Ω to reduce their output. Under normal circumstances all the speakers would have equal impedances to the output of the amplifier.

The final pseudo-circuit called the **Hafler** system, uses four speakers, two at the front and two at the rear as in true quadrophony. By careful selection of resistor values as shown, the ambience can be split into rear left and rear right ambience using the resistor values given. The amount of ambience can be adjusted but at the expense of the power to the main front speakers. The polarity of the speakers is important so that the phasing is correct; simple tests with a small battery will show the polarity of each speaker if it is not marked on the speaker, the cone moving in or out as appropriate. For correct phasing both speaker cones should move in and out in unison. The Hafler circuit resembles the simple circuit when full ambience is required, R_2 being open-circuit, the impedance of each speaker once again being equal to the amplifier impedance. Watch the power rating of the resistors once again; an 8 W amplifier delivers a current of 1 A to a 8 Ω system, so a series 8 Ω resistor must be rated at 8 W.

Further audio designs suitable for car audio boosters can be found in *Electronic Projects 3*.

4.2 Car Aerial Amplifier

The simple circuit, about 250 mm square, in figure 4.7 will boost all radio transmission signals up to VHF by about 10 and so greatly improve the signal-to-noise ratio of the aerial signal to the radio. The circuit consumes only about 2 mA and so can either use a small battery or be wired permanently into the car radio itself, close to the aerial socket. Since this is a high frequency circuit, you are advised to house the circuit in an earthed metal case and as near to the car radio as possible.

4.3 Rev Counter

A 1 mA meter on the dashboard can be calibrated to read 10 000 rpm on full scale deflection using the circuit of figure 4.8. Ignition pulses are detected with a fly-lead wrapped round the top lead of the ignition coil (this lead is **not** connected, it is open-ended). These pulses are filtered by C_1 to eliminate any ringing from entering the circuit (see chapter 1) and passed to the amplifier TR1. When no pulse is present TR1 is off and C_2 charges up via R_3 and D2. When TR1 is turned on by an input pulse, C_2 discharges through VR_1 and D1 thereby generating a voltage across the 1 mA meter. The meter deflection is therefore proportional to the rate at which the points open and close, or the speed of the engine in rpm. A second transistor

56 / Projects for the Car and Garage

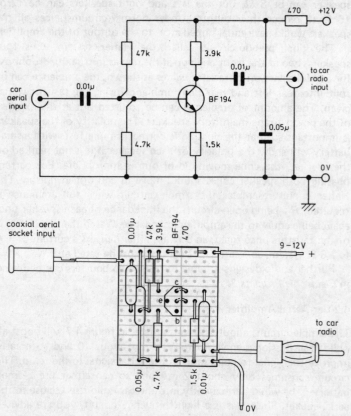

Figure 4.7   Car aerial amplifier.

TR2 is connected as a 9 V Zener diode to prevent battery changes affecting the reading, R_5 being used to calibrate the meter as follows. Obtain a signal generator and use it to calibrate this unit by connecting a 10 V squarewave at the following frequencies via a 1 nF capacitor to the input:

Engine rpm	Pulses per minute	Hz
500	2000	33.3
2000	8000	133.3
5000	20 000	333.3
9000	36 000	600

Accessory Circuits / 57

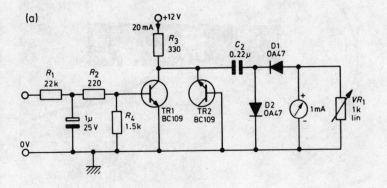

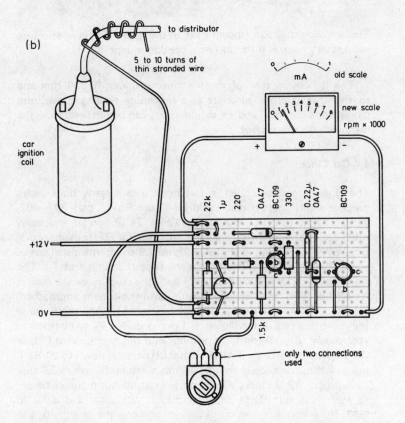

Figure 4.8 Rev counter. (a) Circuit diagram. (b) Layout.

58 / Projects for the Car and Garage

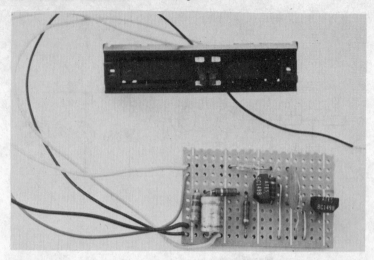

The Veroboard circuit, shown next to the slider. You can, of course use a rotary control if this proves more convenient.

The maximum rpm of most engines is under 10 000 rpm and so this should prove adequate as a maximum reading. Speed/rpm conversions for four- and six-cylinder cars can be obtained from the manufacturer's handbook.

4.4 Car Clock

The clock circuit described in this book uses a ready built multi-purpose multifunction clock sold by Radio Spares part 307—402, with 18 mm (0.7 in.) characters, 12 or 24 hr format and many other facilities which are originally intended for 50 Hz mains operation. When connected as shown in figure 4.9 an internal oscillator of about 20 Hz, set by VR_1, flashes the time at intervals of 1 s. The time can be set by reducing VR_1/R_2 until the clock reads the correct time (slide VR_1 to a very low value until the time is approached, then slide back to mid-track and open-circuit S1). Adjust VR_1 until the repetition rate of the display is 1 per second. VR_1 can be marked accordingly. Crystal control is possible with this circuit using TTL or CMOS divisions from a 1 MHz or 100 kHz crystal down to 20 Hz — this is left to the experimenter. The unit measures 84 mm x 38 mm.

Analogue 12 V clocks are of course available but these are harder to read. This extra large digital display is very clear and quick to read. Note that a reduction of R_1 will increase the brightness, but reduce the life of the display if run continuously.

Accessory Circuits / 59

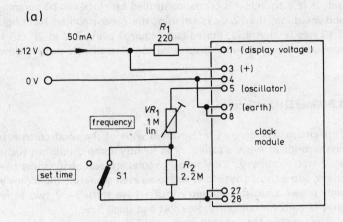

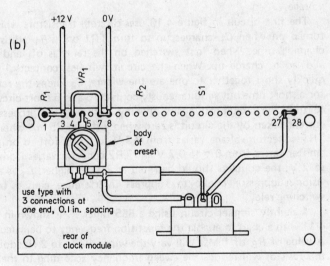

Figure 4.9   Car clock. (a) Circuit diagram. (b) Layout.

A much cheaper form of display is now available in the form of a cheap LED wrist-watch which can be mounted on the car dashboard (remove the straps of course). Of course, the digits are small but for only a few pounds this crystal-controlled car clock can be purchased and wired into the 12 V system using the Zener stabiliser from figure 4.13 (see section 4.7), the voltage required being 3 V at 20 mA for constant display with the display button permanently secured down.

4.5 Wiper Delay Circuit

This circuit will give an intermittent wipe of the windscreen every few seconds in foggy weather, or in slightly damp conditions such as on a wet motorway. Very many circuits are published, some using relays, some using thyristors, and some even operating the motor via high power transistor circuitry, but I have settled for two simple relay circuits for reliability, low cost and small size.

The disadvantage of thyristors used for this purpose is that they need additional circuits to turn them off. Thyristors latch 'on' and do **not** turn off until the anode voltage goes lower than the cathode voltage.

The first circuit in figure 4.10 uses discrete transistors which are turned on when C_1 charges up to turn TR1 on; VR_1 adjusts the charging time. When first switched on the relay is off and C_1 is allowed to charge up. When the circuit switches, contacts 1 and 2 quickly short together to operate the wipers, C_2 keeps the relay on for a short time but simultaneously contacts 5 and 6 short-circuit C_1 to turn the circuit off again, thereby restarting the process. The current taken by the circuit is zero when 'off' and 25 mA when 'on'. TR1 collector voltage varies from 2.3 V to 0.5 V (off to on), TR1 emitter varies from 0 V to 0.7 V and TR2 collector varies from 11 V to 2 V; the contacts should be rated at 2 A or higher. L_1 is a television suppressor choke to suppress interference caused by the switching relay.

A slightly simpler circuit using a 555 timer IC is shown in figure 4.11 with a chart to enable the repetition frequency to be calculated. A value of R_B of 1 MΩ will vary the wiper sweep to 2 Hz; doubling this resistor will reduce the sweep frequency according to the given formula. Only one set of closing contacts is required here and the sizes of the capacitors are smaller. These two circuits merely place a brief short-circuit across the windscreen wiper switch and with modern self-parking wiper motors, **one sweep across the screen and back** is initiated. The unit can therefore be placed near to the wiper switch with the supply leads connected to chassis (0 V) and the input +12 V feed to the wiper switch. On most cars three leads go to the wiper switch, one from 12 V and two to the self-parking switch

Accessory Circuits / 61

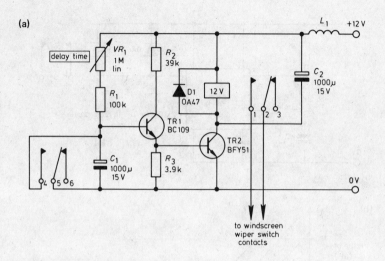

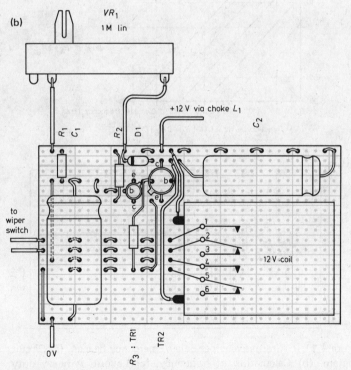

Figure 4.10 Wiper delay. (a) Circuit diagram. (b) Layout.

62 / Projects for the Car and Garage

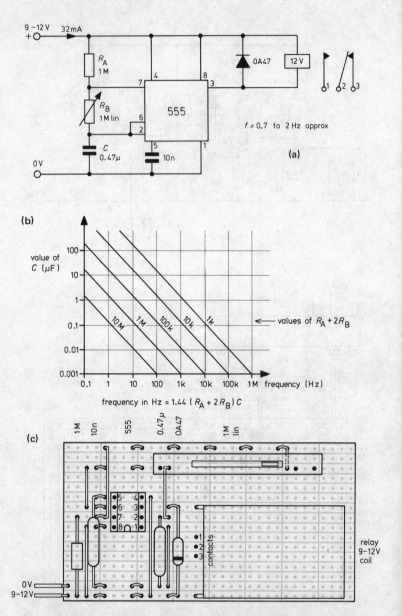

Figure 4.11 Alternative wiper delay/emergency beacon. (a) Circuit diagram. (b) Calculation of frequency. (c) Layout. A heavy duty relay is shown. A Reed relay for the beacon is more compact.

of the wiper motor. The relay contacts are placed between +12 V and the corresponding contact normally made to the motor to start it sweeping across — consult the car wiring diagram if necessary. The chart shows how other time intervals can be obtained by varying the capacitor C or the resistors $R_A + 2R_B$ accordingly, the horizontal scale going from 0.1 Hz (one cycle in 10 s) to 1 MHz (one cycle in 10^{-6} s).

4.6 Emergency Beacon

The same circuit as in figure 4.11 doubles as an emergency flashing beacon of a type similar to the 'roadworks' beacons where a large yellow light flashes at a rate of about 1 Hz. A suggested construction is seen in figure 4.12, built to suit the type of battery chosen, 9 V or 2 x 6 V. The lamp voltage will also depend on the chosen voltage; the wattage can be high, up to 50 W, if 4 A relay contacts are used, although the dry battery will not last for very long. The roadworks beacons use a thermally operated internal switch inside the lamp bulb, along similar lines to flashing Christmas tree light bulbs. These lamps can be purchased from a good electrical supplier thereby saving the need for the 555 circuit, but if you have these few simple components around you, this circuit can be constructed in half an hour. One very efficient roadside light works from a xenon tube using the ignition timing light circuit of chapter 1 and a thyristor to cause it to flash on or off. The method adopted for this flashing xenon is outlined in figure 4.12b where the thyristor gate is triggered when the capacitor C_1 has charged enough to cause the thyristor to fire. When this happens, the high voltage energy in the capacitor C_2 is conducted to the xenon tube causing it to flash. There is no end to flashing beacon circuits.

A ready-made printed circuit board for the d.c. to d.c. converter and for the complete ignition circuit is available from Maplin Electronic Supplies Ltd — see p. 108.

4.7 Car Appliance Supplies

Many electronic devices such as cassette recorders, radios, shavers, vacuum cleaners and other low voltage dc devices do not always run from a 12 V supply. They may be run from the car battery, however, using the simple Zener diode stabilising circuit of figure 4.13. The 12 V supply is tapped from the ignition switch as before, brought

64 / Projects for the Car and Garage

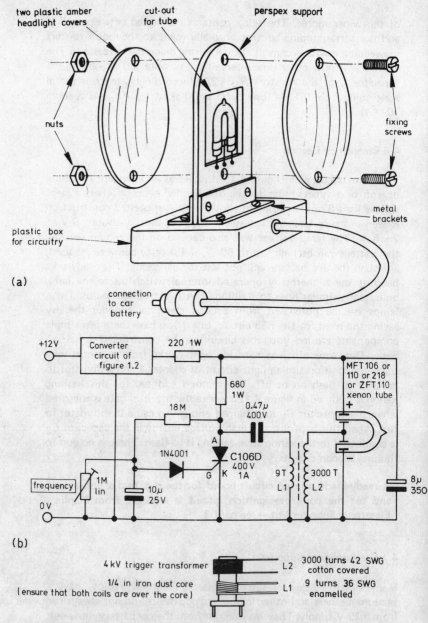

Figure 4.12 Emergency flashing beacon. (a) Layout. (b) Circuit diagram.

Accessory Circuits / 65

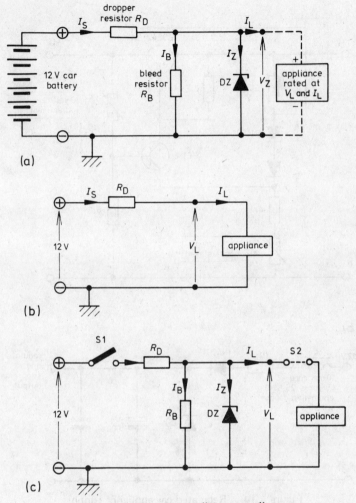

Figure 4.13　Car appliance supplies.

out to a dashboard socket via a fuse (or the cigarette lighter socket may be used if this is provided).

The circuit comprises a dropper resistor R_D which drops the voltage from 12 V to V_L as in figure 4.13b, the value of R_D being calculated from V_{drop}/I_L Ω. The disadvantage of this circuit is the inevitable variation of voltage at the appliance owing to variations of car battery voltage and current into the appliance; a typical cassette

66 / Projects for the Car and Garage

(a)

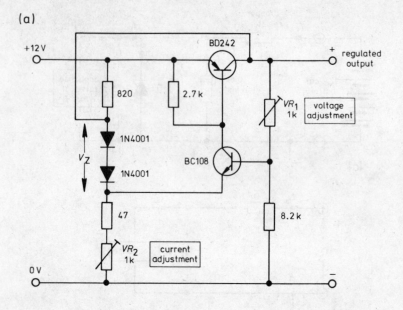

(b)

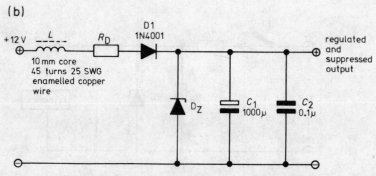

Figure 4.14   Regulated car appliance supply.

player consumes 100 mA when rewinding but only 50 mA when playing.

The Zener diode is added to limit the upper voltage level to that of the appliance with R_D being calculated to assume the **highest** current into the appliance. A bleed resistor is added (R_B) to safeguard the Zener diode when the appliance is disconnected, the Zener then having to conduct **all** of the current and possible damage will result. The usual Zener voltages are 2.7, 3.3, 3.9, 4.3, 4.7, 5.1, 5.6, 6.2, 6.8, 7.5, 8.2, 9.1, 10 and 12 V. Zener diodes are normally rated at 400 mW or 1.3 W for normal use, and the 400 mW can sink 33 mA if the appliance is disconnected from the 12 V supply; the 1.3 W can sink about 100 mA.

R_B is normally a 0.5 W resistor, 560 Ω will sink 50 mA to chassis and finally R_D, for a 7.5 V cassette recorder whose maximum current consumption is 100 mA, calculates to 4.5/0.1 = 45 Ω, 47 Ω being the nearest value.

The position of the on/off switch is very important so as to avoid excess current in the Zener; S2 in the appliance should not be used to switch on and off. Instead a switch at the input S1 should be inserted so that the entire circuit can be disconnected. This is a safety precaution which is more important than one might realise, because the Zener becomes rather warm when constant maximum current is passing through, and a fire could result. Apart from this the battery would be constantly delivering this current of 50 to 100 mA and a poor battery would soon fall flat. This small circuit could surely be inserted into the appliance itself using a little ingenuity and the switch moved from position S2 to S1 in figure 4.13c.

A more reliable circuit is illustrated in figure 4.14a where a *pnp* power transistor which is capable of delivering 4 A is placed as a series regulator between the car battery and the appliance. VR_1 adjusts the output voltage and VR_2 adjusts the output current to suit the appliance voltage and current ratings. A voltmeter can be used to set the voltage via VR_1; connection of the appliance will set VR_2 so that the regulator cuts off at currents above that set by VR_2. The two 1N4148 diodes act as a 1.2 V Zener diode; the outputs will be increased if a real Zener diode of higher Zener voltage is used. If the appliance only consumes about 50 mA, or less, then a small *pnp* general purpose transistor can be substituted for the BD242. Alternatively if higher currents are required a *pnp* 2N3055 can be substituted and a heat sink carefully connected.

One final point concerning appliance supplies concerns suppression from the ignition and car electrical circuits. Figure 4.14b shows how an inductor and diode D1 can be placed into any of the given circuits to supress high frequency interference. Another feature of this circuit is C_2, a small ceramic capacitor which is often placed across a large electrolytic capacitor to cancel out the internal inductance, thereby avoiding low frequency oscillations in the accompanying circuits. If this form of low frequency oscillation is heard (sometimes referred to as 'motorboating') then the addition of C_2 should solve the problem.

4.8 Ammeter and Voltmeter Connections

Once upon a time cars were built like cars and accessories were included as standard fixtures to help the motorist with engine-fault diagnosis as he drove the car along — I refer to the ammeter, the voltmeter, the several gauges and indicator lights and other standard units now added at considerable extra cost. In the early days of motoring there was more likelihood of trouble with the engine and battery but servicing costs were minimal; today the car components such as car battery and ignition points and plugs are more reliable but servicing costs have risen so steeply that many motorists cannot afford the cost of regular servicing. In addition, they often cannot afford the **time** to service the car themselves and so the engine and battery deteriorate until one cold wintry day the starter is turned with a ... chug ... chug ... and it gives up.

A centre-zero meter which can be purchased from any good supplier of surplus stock can be installed as a car ammeter by connecting **across** the battery lead and **not** in the dangerous position in **series** with the battery output; the correct method of connection is seen in figure 4.15. A 100 μA meter will give a good indication of positive current leaving the battery or negative current entering the battery while charging and so a quick check on battery condition is provided. As each appliance is switched on the positive current will increase, particularly the starter motor; as the engine is switched on the negative current should overtake the positive current giving an overall negative reading. If it does not, you have problems. The exact current taken is unimportant, only the direction of travel is required to show the charge/discharge state of the battery or dynamo. The leads to the meter should be of normal flex used for car wiring and preferably as short as possible.

The voltage of the battery can also be accurately measured with a 5 V voltmeter placed across the car battery as seen in figure 4.15. This is sometimes called an extended-range voltmeter since a 10 V Zener rejects the unwanted 0 to 10 V readings and leaves just the range required for normal use. If a 5 V meter is not available a meter of between 5 and 10 V will be suitable or a 1 mA meter connected with the appropriate series multiplier resistor R_M. The value of R_M calculates to $5000 - R_1$ where R_1 is the internal resistance of the 1 mA meter.

The ambitious constructor can use just **one** 1 mA meter on the dashboard to read battery voltage (using series R_M), **or** battery current (if centre zero), or dwell angle (figure 1.8), or rpm (figure 4.8), **or** acceleration/deceleration (chapter 5) by careful switching and wiring.

Accessory Circuits / 69

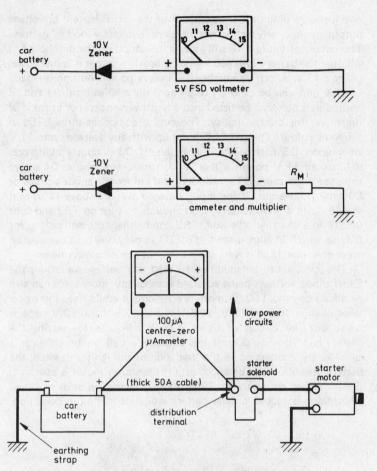

Figure 4.15 Ammeter/voltmeter.

In order that the centre-zero ammeter reads charge and discharge of the battery and does not register the excessive starter motor current, move the connection to **all** low power car circuits, including lights, from the battery positive terminal to the other end of the 50 A cable, on the starter motor solenoid, as seen in figure 4.15.

4.9 Battery Condition Tester

There are many circuits and devices available which will test the specific gravity of the acid, test the A hr capacity of the battery or give just a reading on the dashboard of the actual current or voltage output so that early action can be taken to detect a 'dodgy' battery. The circuit of figure 4.16 will give an indication on the dashboard of whether the battery is in good running order, whether it is just about giving a 12 V output or whether it is in very poor condition on load.

This unit can be built on to a board only 30 mm square and, if housed in a box, can be fitted into any convenient corner in front of the driver for clear visibility. The circuit comprises three LEDs of different colour. The red LED lights up with any voltage up to 11 V whereupon D3 comes into the action. If D3 conducts at its predetermined 11 V potential then TR1 turns on to light up D4 and at the same time cause D2 to conduct and reduce the anode voltage of D1 until it goes out. If the input voltage now goes above 12 V, as it should do in a good battery, D6 conducts to turn on TR2 and light up D7 to show that all's well. TR2 conducting automatically turns D3 out which in turn is turning off D1 so only one light can ever be on at one time. If all is well, the green light should always be on.

This circuit can be modified to read any voltage variation, the Zener diode voltages being adjusted accordingly; more steps can also be added beyond TR2. A more precise circuit could have five or six steps reading 10.5 V, 11 V, 11.5 V, 12 V, 12.5 V and 13 V respectively, but the circuit of figure 4.16 is adequate for testing the battery condition quickly. If this circuit is on all the time the car is running, by connecting to the 'hot' side of the ignition switch, the battery condition can be tested as: (1) the starter motor is operated; (2) the lights are operated; (3) the car is revved up or as any load whatsoever is placed on the battery which, if in tip-top condition,

The positions of the three LEDs can be changed — it may prove more convenient for mounting to solder the LEDs to the *back* of the board, or even put them on flying leads.

Accessory Circuits / 71

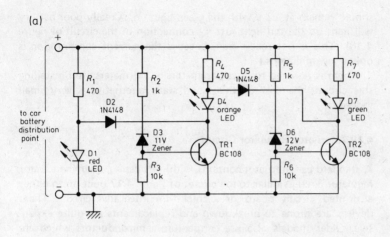

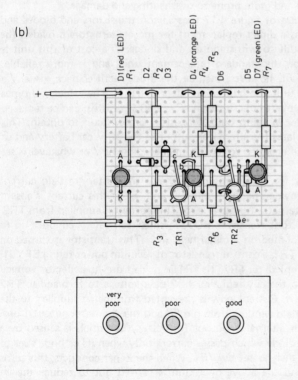

Figure 4.16 Battery condition tester. (a) Circuit diagram. (b) Layout.

should remain at 12 V with the green light on. A really poor battery will light up the red light just by connection of the circuit of figure 4.16! (This is not really true, because the current consumption is only a few milliamps.)

There is no need to install ammeters or voltmeters after installing this circuit; the cost of this 'solid state' alternative is very small indeed.

4.10 Alternator Regulator Circuit

A standard car alternator contains, within its casing, a semiconductor regulator circuit similar to the circuit of figure 4.17 built on to either a printed circuit board or a thick-film integrated circuit. These devices are prone to breakdown and replacements are quite expensive. Older models also use germanium semiconductors which are less reliable and more prone to overheating and damage.

The circuit of figure 4.17 uses silicon transistors and diodes and will serve as a direct replacement for most alternators provided the circuit is built to withstand oil and sludge. The cost of this unit is far less than the standard replacement units and is more reliable. This does not, however, include the diode rectifier pack, usually a three-phase unit also built as an integrated unit. The values and types of component are not critical and R_1, a 50 Ω resistor, can be replaced by a 12 V 2 W lamp bulb if resistors are difficult to obtain. This circuit regulates the alternator output to the 12 V car battery and so prevents the charging voltage from exceeding 12 V or whatever is set by VR_1.

In order for the alternator to charge the battery, a field current must be provided between pins '−' and 'F'. If this current is absent then the alternator output is zero. This current is supplied from TR2, a power output transistor BD131 which can safely conduct up to 6 A but mounted on a metal heat sink. This transistor is turned on and off by TR1, a control transistor of medium power rating BFY51, directly coupled to TR2. If TR1 is turned on, its collector voltage falls to zero thereby reducing TR2 base voltage to zero and so TR2 is turned off. Current now is prevented from being supplied to the alternator field winding from pin 'F' and the alternator output drops.

A rise in alternator output (or battery potential) is sensed by a 9 V Zener diode which passes current only when its cathode rises to 12 V or higher, as set by VR_1. When the Zener conducts, this turns on TR1, and the above operation is carried out to reduce the increased alternator output back to 12 V. VR_1 and R_3 form a potential divider to TR1 base, R_2 limits the base current and stabilises the

Accessory Circuits / 73

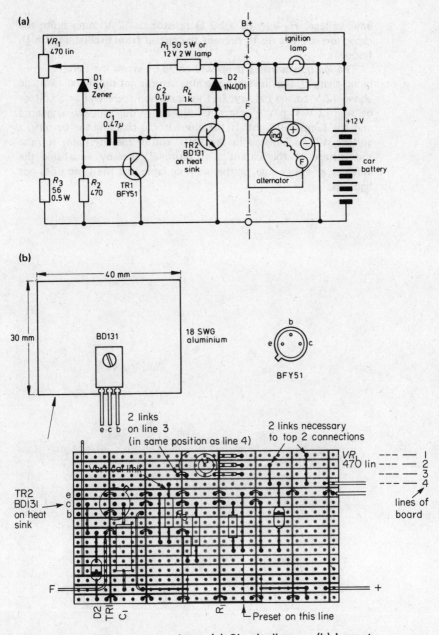

Figure 4.17 Alternator regulator. (a) Circuit diagram. (b) Layout.

base voltage. R_1 can be a 50 Ω resistor or 12 V lamp bulb; the capacitors are merely to prevent the circuit from oscillating at high frequency.

The *ignition warning lamp* will come on when the battery is not being charged, that is, when the alternator is not supplying a voltage above 12 V to the battery. In this condition there will be a voltage below +12 V at pin '+', current will flow through the warning lamp and the lamp will light. If the alternator *is* charging the battery, a voltage will be induced in the 'IND' coil of the alternator to raise the voltage at the circuit's '+' terminal, thereby equalising the voltages at either side of the warning lamp and the lamp does not light up.

5
Garage and Test Gear

This chapter describes a few circuits which are not necessarily connected to the 12 V car battery but may be used in the garage or car as aids to servicing or prevention of a nasty parking fine from a person in yellow and black. For this reason these circuits do not all require 12 V supplies although they may be adapted for connection to the car battery.

5.1 Battery Charger

As with car ignition circuits and windscreen wiper delay circuits, there is an abundance of charger circuits found in magazines and books, and every one claims to be a winner. I have chosen a circuit which proves to use the minimum of components, is therefore low-priced and is safe to use; it can be left alone without worry of overheating or trouble.

Before the circuit is described it is essential that the reader fully understands the nature of the beast being charged, namely the lead—acid 12 V accumulator. The internal resistance is very low indeed, unlike a dry battery which has a high internal resistance equivalent to a series resistance in the output lead. A car battery can deliver

300 A or so when required. The charger must feed several amps into the battery to reverse the internal chemical process.

A car battery must be in tip-top condition at all times if it is to be relied on to supply current to the starter motor when required, and to the many other circuits listed in the introduction to this book; the number of devices increases daily in new cars and so the battery becomes more vital as a key source of energy. If not kept fully charged for a length of time (several weeks) a chemical reaction called **sulphation** sets in and a layer of lead sulphate collects on the plates that effectively places series resistance in the output leads and also reduces drastically the capacity of the battery. Sulphation is **not** always reversible and so the onset of this process should always be avoided by keeping the battery fully charged either in the car while making regular runs or by using a charger at regular intervals.

The condition of the battery can be checked by measuring the specific gravity of the acid and topping up with distilled water as necessary. A hydrometer will measure this for you, typical readings being 1.28 for a fully charged battery and 1.11 for a fully discharged battery, both at 21 °C. The condition of a car battery is very temperature dependent, the capacity falling drastically on a freezing day. The plates should never be allowed to get dry, for they disintegrate if not covered in electrolyte. Regular inspection of acid level and of the colours of the positive plates (normally chocolate-brown owing to a layer of lead peroxide) and negative plates (slate grey) is essential. A small amount of lead sulphate often appears on the terminals; this can be reduced by smearing the terminals with vaseline or silicone grease. If this is not corrected the layer of sulphate gets thicker and harder until a leakage path to chassis is set up to discharge the battery quickly.

Regular charging involves charging at a rate of about 4 A for 12 hr or so, or a lower rate for a longer period. Some chargers have a boost rate of 100 A or so which should **never** be used at the 10-hour rate — this will **seriously** damage the battery. The charger circuit of figure 5.1 will charge at 4 A until the battery voltage reaches 13 V at which time the rate falls to less than 1 A to trickle-charge the battery, therefore maintaining it in a fully charged condition. **Never overcharge a battery**; the 4 A rate for several days would constitute overcharging. The charger has the following facilities

(1) Output current limited to 4 A for a battery in **any** condition.

(2) Current remains at 4 A until charged, followed by an indefinite trickle-charge.

(3) Over-charging is impossible.

(4) The charging rate can be adjusted via VR_1.

Garage and Test Gear / 77

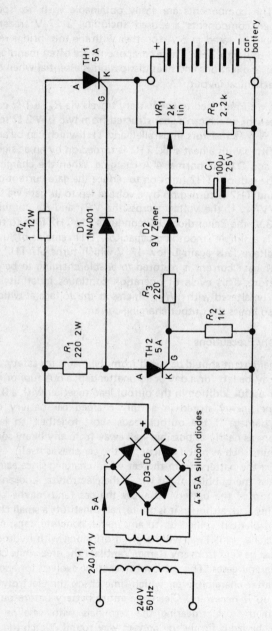

Figure 5.1 Battery charger, circuit diagram.

(5) The components are easily obtainable with no 'specials'; only sixteen components are used, including a 17 V transformer which should be rated at 5 A together with the four bridge rectifier diodes and thyristor TH1. These components are often manufactured for charger applications and heat dissipation is essential when designing the mechanical layout.

The main current feed to the battery travels via R_4, a 1 Ω current-limiting resistor which can be constructed from two 6 W 2 Ω resistors or four 3 W 4 Ω resistors in parallel, and TH1 which can be any 5 A thyristor fitted with a heat sink. TH1 is turned on by an ac signal fed from R_1 and D1 for normal 4 A charging. When the charging rate needs to be reduced TH2 turns on to reduce the gate current of TH1 to zero, and TH2 is turned on by a voltage fed to its gate via D2 and R_3 from VR_1. If the voltage across the VR_1 and R_5 combination reaches 13 V, the Zener diode D2 conducts and TH2 fires to turn off TH1; C_1 is a simple smoothing capacitor. TH1 remains off until the output voltage falls again below 13 V which turns off TH2 whereupon TH1 gate current is restored to enable charging to be carried out as before. This cycle of operation continues indefinitely. The cycle can be altered with VR_1, which sets the voltage at which TH2 fires and so limits the output charging current.

5.1.1 Safety Precautions

The charger circuit should be fused from the mains for safety reasons since it may be left for a day or so unattended; a 5 A fuse or cut-out is another useful addition in the output lead together with a 0 to 5 A meter. The energy stored in a fully charged car battery can do immense damage if the output leads short together, so leave all connections as neatly as possible well away from any likely source of short-circuits such as metal benches or the car chassis itself.

Remove the battery from the car when charging since acid could spill over as the bubbles of gas rise in the electrolyte. Loosen the six caps on top of the battery to allow the gas (and maybe acid) to escape. This acid, although it is dilute, can constitute a small chemical processing plant if spilled on to any metal, concrete, rags, clothing and especially skin! Heat will be given off along with hydrogen gas which must be kept from any flame. **Ventilate** the area while charging to remove any gases. This gas continues to be evolved for about half an hour after charging, after which time check the electrolyte level and top up if necessary. Clean the entire battery before putting it back into the car, smearing the terminals with vaseline before replacing the leads firmly **the correct way round**. Watch the small + and − marks on the battery, close to the two corresponding terminals.

5.1.2 Layout

The layout for the charger circuit is seen in figure 5.2, except for the 17 V transformer, fuses and switches which can be mounted to suit the box provided. This box should be metal, and adequately earthed and ventilated to allow any heat to escape. The bridge rectifier diodes are shown as separate diodes but encapsulated versions are available if size is limited. Similarly the 1 Ω resistor can have one of many alternative forms.

The layout also indicates those wires which will carry up to 5 A of current continuously; these should be of stout 5 to 10 A wire (18 swg or equivalent) and careful planning will keep these wires as short as possible. When mounting the diodes and thyristors remember that the metal chassis is not always to be connected to the metal parts of the device; insulating washers may be needed.

5.1.3 Testing

Firstly connect up the unit of figure 5.2 without any battery and check that the output voltage varies from 12 to 13 V as VR_1 is adjusted. Then connect a fully charged battery and check that 4 A is the upper limit of current that flows (a 5 A meter is necessary of course); adjust R_4 as necessary until the upper limit is correct. Adjust VR_1 and check that the circuit switches between full charge and trickle-charge as TH2 turns on and off. The unit may now be

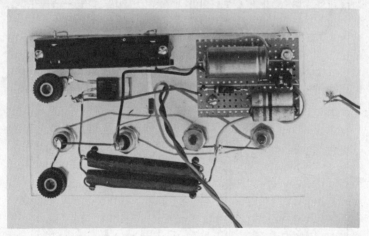

This unit is based on Perspex so that you can see the wiring under the chassis. The working unit has a metal chassis, which acts as a heat sink, but it is essential to use insulating washer sets to isolate the diodes and the thyristor from the metal base.

80 / Projects for the Car and Garage

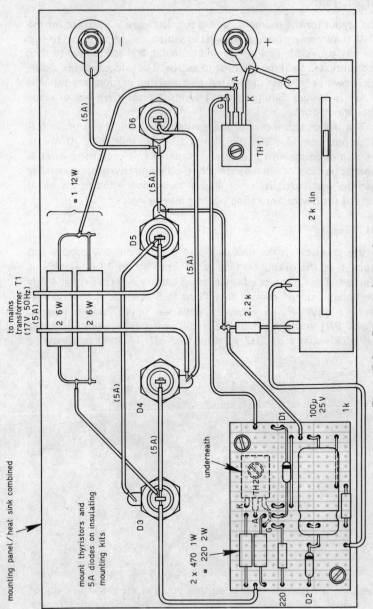

Figure 5.2 Battery charger, layout.

connected to a flat battery with confidence. If trouble is experienced owing to faulty components or faulty wiring, always remove the car battery while checking resistance or voltages. Similarly, remove the mains input while checking resistance — many a meter needle has been wrapped round its end-stop because of thoughtlessness. If it is any consolation to the reader, both my thyristors were faulty in the prototype — the supplier gave me the wrong pin connections. The result was no output at all, but a quick voltage check soon diagnosed the trouble.

5.2 Ice Warning Device

The following circuits use the 741 **operational amplifier**, a small low-priced high-gain IC here used as a **comparator** which will sense whether an input is a few microvolts above or below a preset value to switch on lights or an alarm. The gain is 100 000 and so these few μV of difference are greatly amplified to **switch** the amplifier output from **positive saturation** (that is, almost +9 V) to **negative saturation** (that is, almost −9 V). This does no harm to the circuit.

As a garage ice-alarm (see figure 5.3), the ice probe is situated near to the car so that it warns you to expect antifreeze problems due to the freezing temperatures around. The rest of the circuit may be put in a place where you can see the lights which switch from green to red as freezing point is reached. The circuit makes use of

The Veroboard layout, showing a possible position for the LEDs.

82 / Projects for the Car and Garage

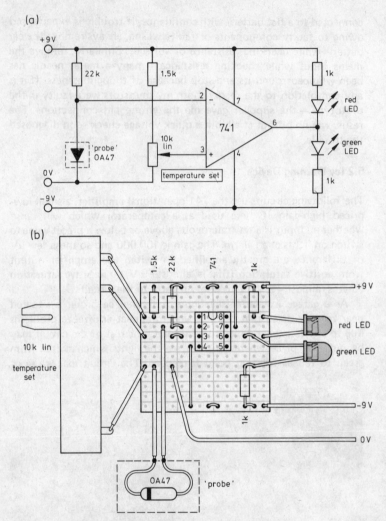

Figure 5.3 Garage ice warning circuit. (a) Circuit diagram. (b) Layout.

the fact that the voltage across a forward biased gold-bonded silicon diode or point-contact germanium diode rises and falls about 2 to 3 mV for every °C change in temperature; types such as the OA91, OA47, 1N914 and 1S914 are suitable. These diodes all work well at freezing point. The 10 kΩ control is set to switch the circuit at 0 °C by placing the diode in a fridge and adjusting until switching just takes place when the lamps will switch. The unit consumes 10 mA (positive) and 4 mA (negative) and so will run for months from 9 V dry batteries. The wiper of the 10 kΩ control extends from 0 to +8 V, pin 6 switches between +8 V and −8 V. An audio alarm can be substituted in place of the lights if wanted; the 741 is capable of driving those described in this book. When wiring the LED devices watch the polarity of the leads, since different manufacturers use different markings. This unit can be used to detect low or high temperatures in freezers, fridges, boilers, car engines, car interiors or indeed any place where monitoring is needed.

5.3 Car Ice Alarm

One minor modification to this circuit (see figure 5.4) enables the circuit of figure 5.3 to be used on a car to sense that ice is on the road ahead. An additional 47 kΩ resistor is added to deliver the requisite supplies to the 741 from the single +12 V car battery. The diode probe should be protected from the weather and dirt by encapsulation in plastic or epoxy resin and positioned at the front of the car, near to the bumper.

5.4 Gauge Alarm Circuits

Yet another use can be made of the 741 comparator circuit to raise an alarm in the car if the fuel is very low **or** if the temperature of the engine is very high (temperature gauges are not always fitted, however). Point A in figure 5.4 is connected **either** to the petrol gauge **or** the temperature gauge; both are normally hot-wire ammeter instruments whose voltage changes can be sensed by the 741 circuit shown. LED displays or audio alarms can be used at point D, points B and C need not be connected if the instrument sender unit is earthed or connected to +12 V as shown. To set up the circuit the 10 kΩ control is again adjusted at danger level until the circuit switches.

5.5 Parking Meter Reminder

A parking fine these days costs the motorist from £6 to £10. This circuit can be constructed for little over £1, or less if a few spare components are available; the circuit is shown in figure 5.5. The

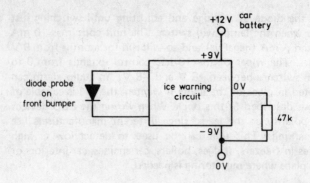
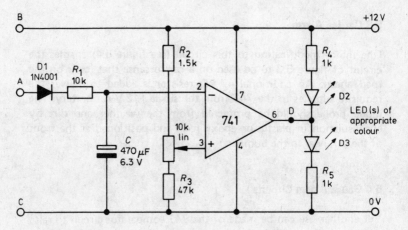
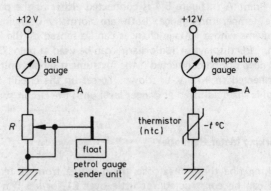

Figure 5.4 Car ice alarm/gauge alarm.

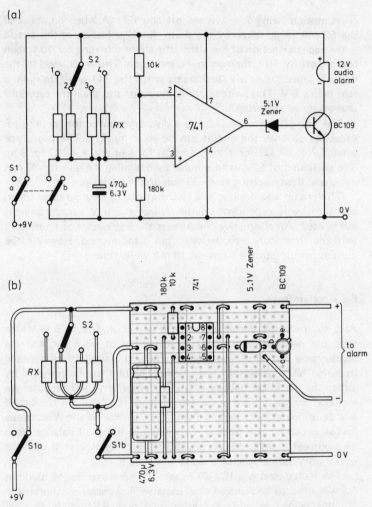

Figure 5.5 Parking meter reminder. (a) Circuit diagram. (b) Layout.

470 µF capacitor charges up, when the unit is switched on, until the 741 comparator switches; the time taken to charge depends on the value of R_X and 470 µF. The voltage on pin 2 is about 8 V and so, when the capacitor charge reaches 8 V, the 741 switches to cause the Zener to conduct and switch on the alarm. The alarm can be either the audio module or the 555 circuit as shown in chapter 3. The Zener diode ensures positive switching of the transistor, current

consumption being 2 mA when off and 60 mA when on, most of the 60 mA being taken by the alarm. Another feature of this circuit is the **repeater** action of the alarm, the alarm sounding for 10 s, then turning off for 10 s, then on again, and so on. This is achieved by the 470 μF capacitor slowly discharging when the 741 is off until pin 3 falls below 8 V. This switches the comparator and again the capacitor charges up *ad infinitum*.

The on—off switch doubles as a discharge device for the 470 μF capacitor so that the circuit can be reset quickly. The prototype used R_X as 27 kΩ for 1 min, 1.5 MΩ for 1 hr and 4.7 MΩ for 3 hr. The switching of S2 can be avoided by designing the unit for 30 min delay and then resetting every 30 min until time is up.

This circuit also doubles as a kitchen timer or photographic timer, and the size is dependent on the size of the 9 V battery and the alarm used. An alternative circuit uses the 555 circuit of figure 4.11 with the alarm connected between pin 3 and ground. However, the current consumption is higher at 10 mA while timing.

5.6 Accelerometer

And so the book decelerates to a final circuit (figure 5.6) which tests the **deceleration** or braking efficiency of your car, and the **acceleration** or pull of your engine. The unit is similar to those used by some MOT garages to test brakes. The unit can be powered by either the 12 V car battery or a 9 V dry battery, and facilities are included for calibration under **any** voltage. The basis for the circuit is a dc **bridge** circuit (seen in figure 5.7), similar to the **Wheatstone bridge** circuit, a well-trodden circuit which indicates balance when zero current flows in the meter. If VR_1 or VR_2 is altered then the circuit goes out of balance and the meter reads.

The meter used is a 100 μA meter; a 1 mA meter can be used but R_4 will need to be reduced or eliminated. The meter is calibrated in '*g*', this being the unit of gravitational pull, 9.81 m/s/s. If a car accelerates at a pull of 1 *g* then the resultant pull will be found by combining the forward 1 *g* with its downward 1 *g* pull to constitute a pull at 45° to the horizontal. Similarly a deceleration of 1 *g* gives a pull of 45° in the other direction, as seen in figure 5.6. This angle is measured by the meter by having a heavy weight attached to a potentiometer (VR_2) and using the swinging of the weight to measure the acceleration in *g*. This weight is referred to as the **pendulum** and is constructed from lead as shown. VR_2 is supported vertically to allow the pendulum to swing in either direction, and switch S1

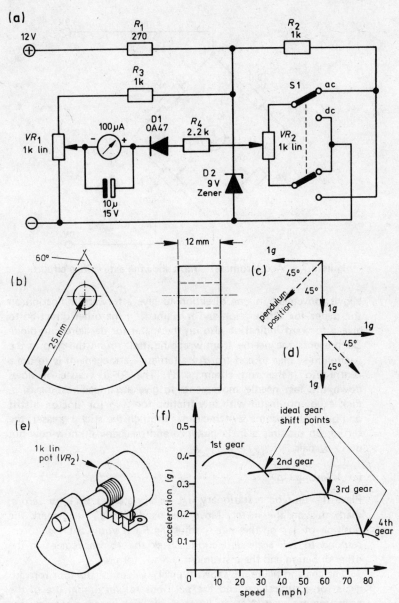

Figure 5.6 Accelerometer. (a) Circuit diagram. (b) Lead pendulum.
(c) 1 *g* acceleration. (d) 1 *g* deceleration. (e) Potentiometer in place.
(f) Graph of acceleration.

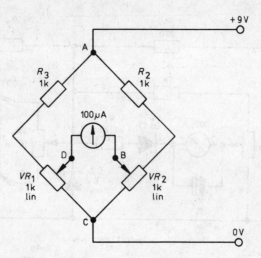

Figure 5.7 Accelerometer circuit showing a dc bridge circuit.

allows movement in one direction to give a forward indication on the meter for acceleration and movement in the other direction to give a forward indication also on the meter for deceleration; diode D1 prevents the needle from wrapping itself round the stop as the pendulum swings to and fro. An alternative arrangement is to use a centre-zero meter and eliminate S1. The 10 μF capacitor slows down the fast needle movements to give a damped reading of g. Pick a potentiometer with **low friction** for VR_2 (or 'doctor' a stiff one by sloshing some switch-cleaner through the sticky grease); the setting up requires a little patience and mechanical know-how but don't despair.

5.6.1 Setting Up

Place the unit on a stationary flat surface and set VR_1 for zero g; mark this on the meter. Move the pendulum 45° and mark the position of 1 g on the meter, adjusting R_4 if you want 1 g to correspond to full scale deflection. Check the 45° movement in the other direction and the instrument is ready for use.

The chart in figure 5.6 shows typical g values for the four forward gears for a typical car and the positions for maximum use of the gears can be seen. You will notice the high speeds necessary for optimum performance — this accounts for the high revs used by racing and rally drivers. Now plot **your own** chart for **your** car and the way **you** drive to check **your** efficiency.

To check the brakes drive at 30 mph and throw out the anchor. The meter should read 0.5 g for drum brakes and 0.7 g for disc brakes. If it reads 10 g you have just driven into the back of a double-decker bus while reading the meter. You should be able to improve your fuel consumption and brake wear by analysing these readings.

Happy motoring!

Appendix I Components Required for Projects

dc/dc Converter (Figure 1.2)

Resistors	33 Ω ¼ or ⅛ W (two) 270 Ω 0.5 W (two)
Capacitors	8 µF 450 V electrolytic 250 µF 40 V electrolytic
Semiconductors	BD711 (two) 1N4004 (four)
Sundries	240 V to 6–0–6 V mains transformer, 6 VA, miniature type

Car Ignition (Figure 1.5)

Resistors	25 Ω 10 W 1 k ¼ or ⅛ W 8.2 k ¼ or ⅛ W 100 k ¼ or ⅛ W

Capacitors	0.22 µF polyester 0.47 µF polypropylene or polyester 1 kV working
Semiconductors	BT109 thyristor, or BT106 or BTY79-400R 1N4001
Sundries	mains neon, any size two non-reversible 3-way sockets, standard or miniature 240 V rating one non-reversible 3-way plug, standard or miniature 240 V rating aluminium or diecast box, size to suit

Timing Light (Figure 1.7)

Resistors	1 M $\frac{1}{4}$ or $\frac{1}{8}$ W
Sundries	Xenon tube ED70 or equivalent

Dwell Meter (Figure 1.8)

Resistors	1 k $\frac{1}{4}$ or $\frac{1}{8}$ W (three) 3.3 k $\frac{1}{4}$ or $\frac{1}{8}$ W 5 k linear pot., preset type
Semiconductors	BC109 (two) 1N4001 (four) 5.1 V Zener diode 400 mW
Sundries	1 mA or 0.1 mA meter, miniature type with access to scale

Ignition Suppression (Figure 1.9)

Resistors	0.05 Ω or 2.5 m of 19 swg copper wire
Capacitors	10 nF low inductance type from car spares supplier 0.5 µF ceramic from car spares supplier 3 µF ceramic from car spares supplier

Ignition Booster (Figure 1.10)

Resistors	1 k $\frac{1}{4}$ or $\frac{1}{8}$ W (two)
Sundries	double pole on–off switch or 12 V relay with change-over contacts Ni–Cd batteries with holder (two), U2 size, 2 A hr capacity such as Radiospares type C, stock number 591-045

Theft Circuits (Figures 2.1 and 2.2)

Sundries	on–off toggle switches with appropriate contacts, see text 12 V relays with contacts rated at appropriate currents mercury switch

Interior Light Extender (Figure 3.1)

Resistors	1.5 k $\frac{1}{4}$ or $\frac{1}{8}$ W 820 Ω $\frac{1}{4}$ or $\frac{1}{8}$ W
Capacitors	470 µF 6.3 V electrolytic
Semiconductors	BC177 BD131 1N4001
Sundries	heat sink, see text

Nightlight (Figure 3.2)

Resistors	100 Ω $\frac{1}{4}$ or $\frac{1}{8}$ W 100 k $\frac{1}{4}$ or $\frac{1}{8}$ W 10 k $\frac{1}{4}$ or $\frac{1}{8}$ W (three)
Capacitors	100 µF, 15 V electrolytic
Semiconductors	BC109 BFY51 OA47 (two)

Sundries	12 V relay with two sets of change-over contacts at appropriate current rating, see text

Lights-on Reminder (Figure 3.3)

Sundries	12 V relay, see text for contact details 12 V audible alarm such as miniature type or figure 3.4.

Alarm Circuit (Figure 3.4)

Resistors	68 Ω ¼ or ⅛ W 4.7 k ¼ or ⅛ W 10 k
Capacitors	0.1 μF ceramic or polyester 10 μF 25 V electrolytic
Semiconductors	555 timer IC and holder
Sundries	miniature loudspeaker, any resistance to suit required volume

Ignition Key In Reminder (Figure 3.5)

Sundries	microswitch, roller type 12 V audible alarm or figure 3.4

Light Monitors (Figure 3.6)

Sundries	fibre-optic cables, terminations, junctions, see text

Lights Failure Monitor (Figure 3.7)

Sundries	dry-reed with contacts rated at 2 A 20 swg enamelled copper wire bobbin for dry-reed 12 V indicator lamp, size to suit, or LED, see text

Audible Turning Indicator (Figure 3.8)

Semiconductors   1N4001   (two)

Sundries   audible alarm or alarm circuit of figure 3.4

Emergency Flasher (Figure 3.9)

Sundries   12 V relay with one on–off contact rated at 1 A
double pole on–off switch

Trailer Flashers (Figure 3.10)

Sundries   12 V relay with single on–off contact (two), contacts 1 A
octal plug and socket

Car Radio Audio Booster (Figure 4.1)

Resistors   2.7 Ω ¼ W
2M log. pot., volume type

Capacitors   0.1 μF polyester
470 μF, 6.3 V electrolytic

Semiconductors   LM380 IC and holder

Sundries   loudspeaker, 8 Ω
heat sink for IC, see text

TBA Booster 5 W (Figure 4.2a)

Resistors   1 Ω ¼ W
56 Ω ¼ or ⅛ W
100 Ω ¼ or ⅛ W
500 k log. pot., volume type

Capacitors	1.5 nF polyester or ceramic
	8.2 nF polyester or ceramic
	0.1 μF polyester
	100 μF, 25 V electrolytic (two)
	470 μF, 6.3 V
	1000 μF, 16 V
Semiconductors	TBA 810 IC
Sundries	loudspeaker, 4–8 Ω
	heat sink, see text

TBA Booster 4 W (Figure 4.2b)

Resistors	1 Ω $\frac{1}{4}$ W
	47 Ω $\frac{1}{4}$ or $\frac{1}{8}$ W
	82 k $\frac{1}{4}$ or $\frac{1}{8}$ W
Capacitors	2.2 nF ceramic or polyester
	0.1 μF polyester
	270 pF silver mica or ceramic
	2.2 μF 25 V electrolytic
	100 μF 25 V
	1000 μF 25 V
Semiconductors	TBA 800 IC
Sundries	loudspeaker 8 Ω
	heat sink, see text

Audio Booster 4 W (Figure 4.3a)

Resistors	1 Ω $\frac{1}{4}$ W
	2.2 Ω $\frac{1}{4}$ or $\frac{1}{8}$ W
	220 Ω $\frac{1}{4}$ or $\frac{1}{8}$ W
Capacitors	100 nF ceramic or polyester
	2.2 μF 15 V electrolytic
	100 μF, 16 V (preferred value)
	470 μF, 6.3 V
	1000 μF, 16 V

Semiconductors	TDA 2002A IC
Sundries	loudspeaker, above 2 Ω heat sink, see text

Audio Booster 16 W (Figure 4.3b)

Resistors	10 k ¼ or ⅛ W 1 M ¼ or ⅛ W (two) 100 k ¼ or ⅛ W
Capacitors	0.1 µF ceramic or polyester 0.47 µF 250 µF, 16 V electrolytic (preferred value)
Semiconductors	LM379 IC
Sundries	loudspeaker, 16 Ω heat sink, see text

Push-Pull Audio Amplifier, 4 W (Figure 4.4)

Resistors	2.7 Ω ¼ W 1 M preset 2 M log. pot., volume type
Capacitors	47 pF ceramic or silver mica 0.1 µF ceramic or polyester
Semiconductors	LM380 IC (two)
Sundries	loudspeaker, 16 Ω heat sink(s) see text

Car Aerial Amplifier (Figure 4.7)

Resistors	1.5 k ¼ or ⅛ W 3.9 k ¼ or ⅛ W 4.7 k ¼ or ⅛ W 47 k ¼ or ⅛ W 470 Ω ¼ or ⅛ W

Capacitors	0.01 µF ceramic or polyester (two) 0.05 µF ceramic or polyester
Semiconductors	BF194
Sundries	coaxial aerial sockets (two) metal box

Rev Counter (Figure 4.8)

Resistors	330 Ω ¼ or ⅛ W 220 Ω ¼ or ⅛ W 1.5 k ¼ or ⅛ W 22 k ¼ or ⅛ W 1 k linear pot., preset type
Capacitors	0.22 µF ceramic or polyester 1 µF, 25 V electrolytic
Semiconductors	BC109 (two) OA47 (two)
Sundries	1 mA meter, see text

Car Clock (Figure 4.9)

Resistors	220 Ω ¼ or ⅛ W 2.2 M 1 M preset
Sundries	Radiospares clock module on–off switch, miniature

Wiper Delay (Figure 4.10)

Resistors	3.9 k ¼ or ⅛ W 39 k 100 k 1 M linear pot., volume type

98 / Projects for the Car and Garage

Capacitors	1000 µF, 15 V electrolytic (two)
Semiconductors	BC109 BFY51 OA47
Sundries	12 V relay with two sets of 2 A contacts suppressor choke from car supplies shop

Emergency Beacon (Figure 4.11)

Resistors	1 M $\frac{1}{4}$ or $\frac{1}{8}$ W 1 M linear pot., preset
Capacitors	10 nF ceramic or polyester 0.47 µF polyester (preferred value)
Semiconductors	555 timer IC and holder OA47
Sundries	12 V relay with contacts to suit rating of lamp(s)

Xenon Flashing Beacon (Figure 4.12b)

Resistors	220 Ω 1 W 680 Ω 1 W 18 M $\frac{1}{4}$ or $\frac{1}{8}$ W 1 M preset pot.
Capacitors	10 µF, 25 V electrolytic 8 µF, 350 V electrolytic 0.47 µF 400 V polyester
Semiconductors	400 V, 1 A thyristor such as BT106 1N4001
Sundries	dc/dc converter circuit of figure 1.2 xenon tube, see text 4 kV trigger transformer, from Maplin or see text

Car Appliance Supplies (Figure 4.13a)

Resistors	R_D and R_B, see text
Semiconductors	Zener diode, see text

Figure 4.14a

Resistors	47 Ω ¼ or ⅛ W 820 Ω ¼ or ⅛ W 8.2 k ¼ or ⅛ W 2.7 k ¼ or ⅛ W 1 k preset pot. (two)
Semiconductors	BD242 BC108 1N4001 (two)

Figure 4.14b

Resistors	R_D, see text
Capacitors	1000 µF, 15 V electrolytic 0.1 µF, 15 V polyester
Semiconductors	1N4001 Zener diode, see text
Sundries	inductor, see text

Ammeter/Voltmeter (Figure 4.15)

Semiconductor	10 V Zener diode, 400 mW or lower
Sundries	5 V voltmeter or ammeter conversion 100 µA centre-zero microammeter

Battery Condition Tester (Figure 4.16)

Resistors	470 Ω ¼ or ⅛ W (three) 1 k ¼ or ⅛ W (two) 10 k ¼ or ⅛ W (two)

Semiconductors	BC108 (two)
	1N4148 (two)
	LED of different colours (three)
	11 V Zener diode, 400 mW
	12 V Zener diode, 400 mW

Alternator Regulator (Figure 4.17)

Resistors	50 Ω 5 W
	56 Ω ½ W
	470 Ω ¼ or ⅛ W
	470 Ω linear pot., preset
Capacitors	0.47 µF polyester or ceramic
	0.1 µF polyester or ceramic
Semiconductors	BFY51
	BD131
	Zener diode 9 V, 400 mW
	1N4001
Sundries	heat sink, see text

Battery Charger (Figure 5.1)

Resistors	1 Ω 12 W
	220 Ω 2 W
	220 Ω ¼ or ⅛ W
	1 k
	2.2 k
	2 k linear pot., preset
Capacitors	100 µF 25 V electrolytic
Semiconductors	400 V 5 A thyristors (two)
	Zener diode 9 V 400 mW
	1N4001
	5 A silicon rectifier diodes (four)
Sundries	240 V/17 V mains transformer, 5 A
	secondary fuses, see text
	switches, see text

Garage Ice Warning Circuit (Figure 5.3)

Resistors	1.5 k ¼ or ⅛ W
	22 k ¼ or ⅛ W
	1 k ¼ or ⅛ W (two)
	10 k linear pot., preset
Semiconductors	741 8-pin IC with holder
	OA47, see text
	LED of different colours (two)

Car Ice Alarm (Figure 5.4)

Resistors	47 k ¼ W
Sundries	garage circuit of figure 5.3

Gauge Alarm (Figure 5.4)

Resistors	1.5 k ¼ or ⅛ W
	10 k ¼ or ⅛ W
	1 k ¼ or ⅛ W (two)
	47 k ¼ or ⅛ W
	10 k linear pot., preset
Capacitors	470 µF 6.3 V electrolytic
Semiconductors	741 8-pin IC
	1N4001
	LED of different colours (two)

Parking Meter Reminder (Figure 5.5)

Resistors	R_X, see text
	10 k ¼ or ⅛ W
	180 k
Capacitors	470 µF, 6.3 V electrolytic
Semiconductors	741 8-pin IC and holder
	5.1 V Zener diode 400 mW
	BC109

Sundries	12 V audible alarm or alarm circuit (555) double pole switch, very low current rating, miniature

Accelerometer (Figure 5.6)

Resistors	270 Ω ¼ or ⅛ W 2.2 k ¼ or ⅛ W 1 k ¼ or ⅛ W (two) 1 k linear pot., preset (two)
Capacitors	10 µF, 15 V electrolytic
Semiconductors	Zener diode 9 V 400 mW OA47
Sundries	double pole change-over switch, very low power, miniature 100 µA centre-zero meter, see text pendulum, see text

NOTE: In all these component lists the current ratings of switches and relay contacts depend on the devices being switched and so the size of the switch or relay will depend on these ratings. For instance, a switch or relay designed to switch car headlamps on and off should be rated at 5 A per 55 W headlamp, and so on.

Appendix II Specifications of Semiconductor Devices

Transistors

Type		Power	I_c	V_{ceo}	h_{FE}	f_T	case
BC107	npn Si	360 mW	100 mA	45	110–450	250 MHz	TO18 or lockfit
BC108	npn Si	360 mW	100 mA	20	110–800	250 MHz	TO18 or lockfit
BC109	npn Si	360 mW	100 mA	20	200–800	250 MHz	TO18 or lockfit
BD131	npn Si	15 W	3 A	45	20 (min.)	60 MHz	TO126
BC177	pnp Si	350 mW	100 mA	−45	500 (typ.)	250 MHz	TO18
BFY51	npn Si	800 mW	1 A	30	40	50 MHz	TO39
2N3055	npn Si power	115 W	15 A	60	20–70	1 MHz	TO3
BF194	npn Si rf	220 mW	30 mA	20 V	115	260 MHz	lockfit
BD242	pnp Si	40 W	3 A	−45	25	3 MHz	TO126
BD711	npn Si power	75 W	12 A	100	25	3 MHz	TO126

Integrated Circuits

555	general purpose timer
LM380	34 dB audio power amplifier in 14 pin dil package
TBA810S	34–40 dB audio power amplifier, see text for pin connections
TBA800	audio power amplifier, see text for pin connections
TDA2002A	4 watt audio amplifier, see text for pin connections
LM379	6 watt stereo audio amplifier, see text for pin connections
741	100 dB operation amplifier in 8-pin dil package

Thyristors

BT106	700 V	1 A	90 °C	$V_{GT} = 3.5$ V	$I_{GT} = 50$ mA
BTY79-400R	400 V	6.4 A	85 °C	$V_{GT} = 3.0$	$I_{GT} = 30$
C106D	400 V	4 A	85 °C	$V_{GT} = 0.8$ V	$I_{GT} = 0.2$ mA

Diodes

1N4001	Si general purpose	1 A	$V_{RRM} = 50$ V	plastic
1N4004	Si general purpose	1 A	$V_{RRM} = 400$ V	plastic
OA47	Ge gold bonded	48 mA	$V_{RRM} = 40$ V	glass, high frequency
1N4148	Si diffused	75 mA	$V_{RRM} = 75$ V	plastic, low power

For high current purposes consult suppliers for required current/voltage rating.

Appendix III Resistor/Capacitor Colour Coding

Resistor Coding

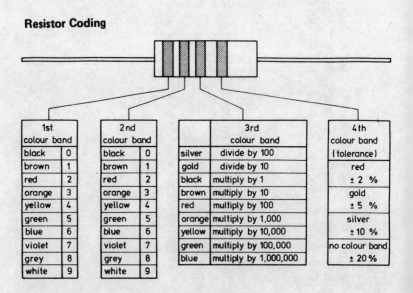

1st colour band	
black	0
brown	1
red	2
orange	3
yellow	4
green	5
blue	6
violet	7
grey	8
white	9

2nd colour band	
black	0
brown	1
red	2
orange	3
yellow	4
green	5
blue	6
violet	7
grey	8
white	9

3rd colour band	
silver	divide by 100
gold	divide by 10
black	multiply by 1
brown	multiply by 10
red	multiply by 100
orange	multiply by 1,000
yellow	multiply by 10,000
green	multiply by 100,000
blue	multiply by 1,000,000

4th colour band (tolerance)	
red	± 2 %
gold	± 5 %
silver	± 10 %
no colour band	± 20 %

Appendix III / 107

Polyester Capacitors

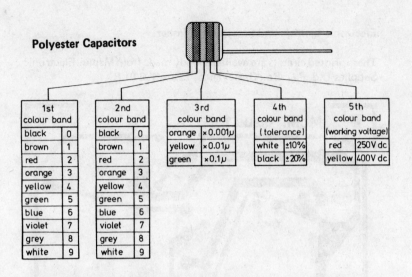

1st colour band		2nd colour band		3rd colour band		4th colour band (tolerance)		5th colour band (working voltage)	
black	0	black	0	orange	×0.001μ	white	±10%	red	250V dc
brown	1	brown	1	yellow	×0.01μ	black	±20%	yellow	400V dc
red	2	red	2	green	×0.1μ				
orange	3	orange	3						
yellow	4	yellow	4						
green	5	green	5						
blue	6	blue	6						
violet	7	violet	7						
grey	8	grey	8						
white	9	white	9						

Standard Decade Values

E24	10	11	12	13	15	16	18	20	22	24	27	30	33	36	39	43	47	51	56	62	68	75	82	91
E12	10		12		15		18		22		27		33		39		47		56		68		82	
E6	10				15				22				33				47				68			

Electronic Ignition and Emergency Flasher

These printed circuits are available ready made from Maplin Electronic Supplies Ltd, P.O. Box 3, Rayleigh, Essex SS6 8LR.

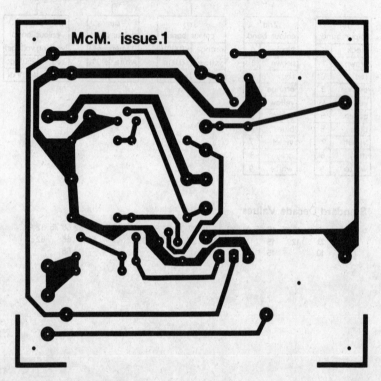

Electronic Ignition

This board is used both for the dc to dc converter and for the complete ignition circuit. Both component layouts follow.

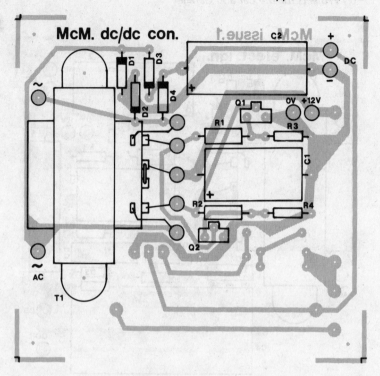

Parts List — dc to dc converter

```
R1, R2          270 Ω std res.
R3, R4          33 Ω min. res.
Q1, Q2          BD711
C1              220 µF 40 V axial
C2              4.7 µF 500 V axial
D1–D4           1N4004
T1              Min. TR6V
Also required:  11 x Veropin 2141
                2 x 4BA screw    ⎫
                2 x 4BA washer   ⎬  for T1
                2 x 4BA nut      ⎭
```

110 / Projects for the Car and Garage

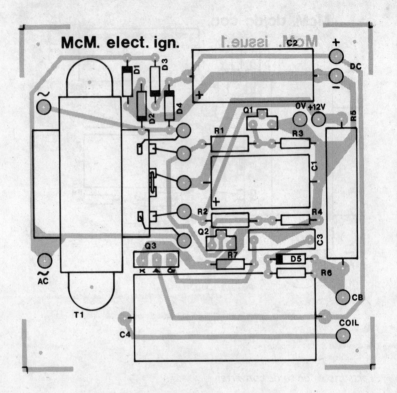

Additional parts for complete ignition system:

R5	22 Ω W wire wound
R6	8.2 kΩ min. res.
R7	1 kΩ min. res.
C3	0.22 μF polyester
C4	0.47 μF 1000 V mixed D.
D5	1N4001
Q3	BT109
Also required:	2 x Veropin 2141

Appendix III / 111

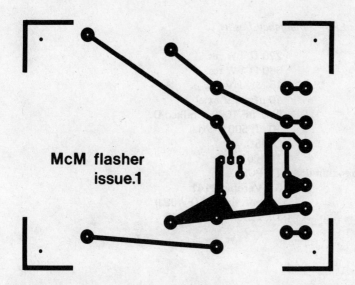

Emergency Flasher

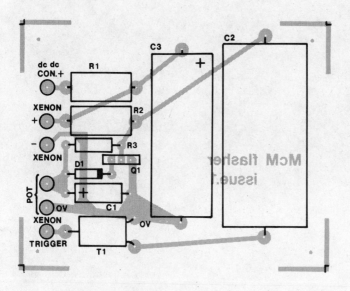

Parts List — emergency flasher

R1	220 Ω 1 W res.
R2	680 Ω 1 W res.
R3	15 MΩ HV res.
C1	10 μF 25 V axial
C2	0.47 μF 1000 V mixed D.
C3	10 μF 500 V axial
Q1	C106D
D1	1N4001
Also required:	1 x PCB
	6 x Veropin 2141
	1 x pot. lin. 1M (FW08J)
	1 x xenon tube NV613

Index

accelerometer 86
accessory circuits 44
accessory supplies 63
aerial amplifier 55
aerial booster 44
alarm, ice 83
alarm circuits 26, 34
alarm module 33
alternator regulator 72
ambience 54
ammeter circuit 68
amplifier, car aerial 55
appliance supplies 63
audible indicator 40
audio boosters 45

battery charger 75
battery circuits 63
battery condition circuit 69
boards
    flasher 108
    ignition 108

boosting ignition 23
bounce 8
bounce suppression 10

capacitance discharge 1
capacitor colour coding 107
capacitor discharge installation 15
car aerial amplifier 55
car clock 58
car ice alarm 83
car radio booster 44
car theft circuits 25
carbon string 21
circuit
    accelerometer 86
    electronic ignition 11
    ice warning 81
    ignition key in 36
circuits
    assessory 44
    immobilising 25
    lighting 29

113

clock 58
component lists 90
contact breakers 2, 17
conventional ignition 3
converter circuit 6
cross-over response 52

dc/dc converter 5
deceleration 86
diode data 105
distributor 2
dwell angle 17
dwell meter 17

EHT leads 21
electronic ignition 8
electronic ignition board 108
emergency beacon 63
emergency flasher 41
emergency flasher board 108
engine tuning 2

fibre-optics 36
flashers
    emergency 41
    trailer 42

garage circuits 75
gauge alarms 83

Hafler system 54

IC data 104
IC radio boosters 45
ice alarm 83
ice warning device 81
ignition booster 23
ignition capacitor 4
ignition circuits 1
ignition coil 2
'ignition key in' circuit 36
ignition principles 2
ignition problems 4
ignition suppression 19

ignition timing light 16
ignition warning lamp 74
immobilise alarms 25
indicator warning sound 40
interior light extender 30

light failure monitor 39
lighting circuits 29
'lights-on' reminder 32

measuring current 68
measuring voltage 68
mercury switch 26
monitor, light failure 39

nightlight 32

operational amplifier 81
over-charging 76

parking meter reminder 83
pendulum 86
power supplies 63
pseudo-stereo 51

quadrophonic 52

radio booster 44
regulating alternator 72
regulation 7
resistor colour coding 106
rev counter 55

safety 78
semiconductor data 103
spark plugs 2
stereo systems 49
sulphation 76
suppression circuits 21

test gear 75
testing dc/dc converter 13
theft circuits 25
thyristor data 105

trailer flashers 42
transistor data 104

voltmeter circuit 68

Wheatstone bridge 86
wiper delay circuit 60

xenon tube 16